YO-AET-978

Progress in Inflammation Research

Series Editor

Prof. Michael J. Parnham PhD
Senior Scientific Advisor
PLIVA Research Institute Ltd.
Prilaz baruna Filipovića 29
HR-10000 Zagreb
Croatia

Advisory Board

G. Z. Feuerstein (Merck Research Laboratories, West Point, PA, USA)
M. Pairet (Boehringer Ingelheim Pharma KG, Biberach a. d. Riss, Germany)
W. van Eden (Universiteit Utrecht, Utrecht, The Netherlands)

Forthcoming titles:

Turning up the Heat on Pain: TRPV1 Receptors in Pain and Inflammation, A.B. Malmberg,
    K.R. Bley (Editors), 2005
NPY Family of Peptides in Immune Disorders, Inflammation, Angiogenesis and Cancer,
    G.Z. Feuerstein, Z. Zukowska (Editors), 2005
Complement and Kidney Disease, P.F. Zipfel (Editor), 2005
*Chemokine Biology: Basic Research and Clinical Application, Volume I: Immunobiology of
    Chemokines*, K. Neote, L.G. Letts, B. Moser (Editors), 2005
*Chemokine Biology: Basic Research and Clinical Application, Volume II: Pathophysiology
    of Chemokines*, K. Neote, L.G. Letts, B. Moser (Editors), 2005
The Hereditary Basis of Rheumatic Diseases, R. Holmdahl (Editor), 2005

(Already published titles see last page.)

Sodium Channels, Pain, and Analgesia

Kevin Coward
Mark D. Baker

Editors

Birkhäuser Verlag
Basel · Boston · Berlin

Editors

Kevin Coward
Department of Pharmacology
University of Oxford
Mansfield Road
Oxford OX1 3QT
UK

Mark D. Baker
Molecular Nociception Group
Department of Biology
Medawar Building
University College London
Gower Street
London WC1E 6BT
UK

Library of Congress Cataloging-in-Publication Data

Sodium channels, pain, and analgesia / Kevin Coward, Mark D. Baker, editors.
    p. ; cm. -- (Progress in inflammation research)
    Includes bibliographical references and index.
    ISBN 3-7643-7062-9 (alk. paper)
    1. Pain. 2. Sodium channels. 3. Analgesics. I. Coward, Kevin, 1969– II. Baker, Mark D., 1960-
III. PIR (Series)

    RB127.S64 2005
    616'.0472--dc22

                                                                2005048132

Bibliographic information published by Die Deutsche Bibliothek
Die Deutsche Bibliothek lists this publication in the Deutsche Nationalbibliografie;
detailed bibliographic data is available in the internet at http://dnb.ddb.de

ISBN-10: 3-7643-7062-9 Birkhäuser Verlag, Basel – Boston – Berlin
ISBN-13: 978-3-7643-7062-6 Birkhäuser Verlag, Basel – Boston – Berlin

© 2005 Birkhäuser Verlag, P.O. Box 133, CH-4010 Basel, Switzerland
Part of Springer Science+Business Media
Printed on acid-free paper produced from chlorine-free pulp. TCF ∞
Cover design: Markus Etterich, Basel
Cover illustration: see page 53. With the friendly permission of Holger Scheib (Department of Structural Biology and
Bioinformatics, University of Geneva and Swiss Institute of Bioinformatics) and Iain McLay (Computational, Analytical
and Structural Sciences, GlaxoSmithKline, Stevenage, Herts, UK).
Printed in Germany
ISBN-10: 3-7643-7062-9
ISBN-13: 978-3-7643-7062-6

9 8 7 6 5 4 3 2 1   www.birkhauser.ch

Contents

List of contributors

Mark D. Baker, Molecular Nociception Group, Department of Biology, University College, London WC1E 6BT, UK; e-mail: mark.baker@ucl.ac.uk

Joel A. Black, Department of Neurology and Center for Neuroscience and Regeneration Research, Yale University School of Medicine, New Haven, CT 06510, and Rehabilitation Research Center, VA Connecticut Healthcare System, West Haven, CT 06516, USA; e-mail: joel.black@yale.edu

James A. Brock, Prince of Wales Medical Research Institute, Barker St, Randwick, Sydney, NSW 2031, Australia; e-mail: j.brock@unsw.edu.au

Fernando Cervero, Anaesthesia Research Unit and Centre for Research on Pain, McGill University, McIntyre Medical Bldg., Room 1207, 3655 Promenade Sir William Osler, Montreal, Quebec H3G 1Y6, Canada;
e-mail: fernando.cervero@mcgill.ca

Jeffrey J. Clare, Gene Expression and Protein Biochemistry Department, Glaxo-SmithKline, Stevenage, Herts, SG1 2NY, UK; e-mail: jeff.j.clare@gsk.com

David Cronk, Ionix Pharmaceuticals Ltd, 418 Cambridge Science Park, Cambridge CB4 0PA, UK

Lodewijk V. Dekker, Ionix Pharmaceuticals Ltd, 418 Cambridge Science Park, Cambridge CB4 0PA, UK; e-mail: ldekker@ionixpharma.com

Sulayman D. Dib-Hajj, Department of Neurology and Center for Neuroscience and Regeneration Research, Yale University School of Medicine, New Haven, CT 06510, and Rehabilitation Research Center, VA Connecticut Healthcare System, West Haven, CT 06516, USA

Michael S. Gold, Department of Biomedical Sciences, University of Maryland Dental School, 666 W. Baltimore St., Room 5-A-12 HHH, Baltimore, MD 21201, USA; e-mail: msg001@dental.umaryland.edu

Bryan C. Hains Department of Neurology and Center for Neuroscience and Regeneration Research, Yale University School of Medicine, New Haven, CT 06510, and Rehabilitation Research Center, VA Connecticut Healthcare System, West Haven, CT 06516, USA

Jennifer M.A. Laird, Bioscience Department, AstraZeneca R & D Montréal, 7171 Frédérick-Banting, Ville Saint-Laurent, Quebec H4S 1Z9, Canada; e-mail: jennifer.laird@astrazeneca@com

Grant D. Nicol, Department of Pharmacology and Toxicology, 635 Barnhill Drive, Indiana University School of Medicine, Indianapolis, IN 46202, USA; e-mail: gnicol@iupui.edu

Kenji Okuse, Wolfson Institute for Biomedical Research, University College London, Gower Street, London WC1E 6BT, UK; present address: London Pain Consortium, Department of Biological Sciences, South Kensington Campus, Imperial College of Science, Technology and Medicine, London SW7 2AZ, UK; e-mail: k.okuse@imperial.ac.uk

Andreas Scholz, Physiologisches Institut, Universität Giessen, Aulweg 129, 35392 Giessen, Germany; e-mail: andreas.scholz@physiologie.med.uni-giessen.de

Stephen G. Waxman Department of Neurology and Center for Neuroscience and Regeneration Research, Yale University School of Medicine, New Haven, CT 06510, and Rehabilitation Research Center, VA Connecticut Healthcare System, West Haven, CT 06516, USA

John N. Wood, Molecular Nociception Group, Biology Department, UCL, Gower Street, London WC1E 6BT, UK; e-mail: j.wood@ucl.ac.uk

Preface

The treatment of chronic pain, for example that resulting from damage or dysfunction of the nervous system, or that associated with cancer, is at present inadequate and pain still represents a serious unmet clinical need. The costs of pain, in terms of personal anguish, finance and in national healthcare costs are enormous. Because sodium channels confer excitability on neurones in nociceptive pathways and exhibit neuronal tissue-specific and injury-regulated expression, their study has become an important branch of pain research, and they form the focus of this book. As well as reviewing why sodium channel subtypes are potentially important drug targets in the treatment of pain, this volume also brings together recent insights into the control of expression, functioning and membrane trafficking of nervous system sodium channels.

A recent previous review of sodium channel function, with particular emphasis on the ways in which aberrant sodium channel behaviour can contribute to nervous system pathophysiology, was based on a Novartis Foundation symposium held in London in 2000, chaired by Stephen Waxman. At that time it had become clear that sodium channels were a group of proteins exhibiting both molecular and functional diversity, and that neuronal hyperexcitability, contributing to such phenomena as chronic pain following nerve injury, might be explained by changes in sodium channel function. This included the selective upregulation and downregulation of expression of different sodium channel genes. The control of sodium channel gene expression in the nervous system following injury has remained very much a hot topic in the intervening years and is an important theme in this book. Evidence has also accrued on the importance of G-protein pathway control of sodium channel function, and the post-translational modification of channel function based on phosphorylation is also discussed in this volume.

The ability to discriminate pharmacologically between sodium channel subtypes, which show substantial sequence homology, is another important theme and one where key developments are expected. The technologies used for screening compounds on sodium channel function are reviewed in this book. Sodium channel subtypes appear to be distributed to specific regions of the axon, and may therefore make highly individual contributions to normal acute noxious sensation. These must include tetrodotoxin-resistant channels, known to be functional in at least some of the smallest peripheral endings. Furthermore, sodium channels are chaperoned to the neuronal membrane and are tethered there by a complex of proteins, contacting both the extracellular matrix and the intracellular cytoskeleton. Many protein–protein interactions must ensure correct channel function and turnover. These interactions can be sodium channel sub-type specific, for example that between p11 and $Na_v1.8$. Thus, certain channel associated molecules might provide additional drug targets in the treatment of pain.

Our understanding of pain transmission and transduction in mammals has been greatly facilitated by the development of sodium channel gene knockout mice. This has allowed us to assign roles to certain sodium channel subtypes that could not be selectively targeted by pharmacological methods, and the endeavour has allowed sodium channel subtypes to be validated as potential future drug targets. The further sophistication of gene knockout technology, developed at least in part to overcome lethality, has been the use of tissue-specific nulls where the activation of a tissue-specific gene promoter can be used to express the bacteriophage cre-recombinase. Finally, the use of tissue-specific inducible nulls is expected to contribute to the future study of sodium channel function. This technology holds out the promise of gene deletion without developmental compensation resulting in a diluted phenotype, and thus may provide the clearest insight possible into the function of genes in subsets of neurones.

This book aims to summarise the current understanding of voltage-gated sodium channels, their association with pain and their potential as targets for the development of novel analgesics. Individual chapters address the potential therapeutic role of voltage-gated sodium channels and their respective roles in neuropathy and nerve injury, brain disorders, visceral pain and dental pain. Further chapters address the role of these molecules in nociceptive endings, the regulation and modulation of sodium channels, channel gating and drug blockade. A specific chapter is devoted to the $Na_v1.8$ channel, viewed by many as an important therapeutic target, and the final chapter discusses current opinion and future direction in sodium channel research.

We wish to thank all the authors who participated in writing this book.

Oxford/London, February 2005
Kevin Coward
Mark D. Baker

Voltage-gated sodium channels and pain associated with nerve injury and neuropathies

Joel A. Black, Bryan C. Hains, Sulayman D. Dib-Hajj and Stephen G. Waxman

Department of Neurology and Center for Neuroscience and Regeneration Research, Yale University School of Medicine, New Haven, CT 06510, USA, and Rehabilitation Research Center, VA Connecticut Healthcare System, West Haven, CT 06516, USA

Introduction

Neuropathies and injury to peripheral nerves produce pathophysiological alterations within sensory neurons that often lead to the development of chronic pain, termed neuropathic pain. Neuropathic pain is generally manifested as an ongoing burning sensation in affected regions and/or by painful reaction to normally innocuous thermal or mechanical stimulation (allodynia), and is often refractory to treatment [1]. Several animal models (e.g., [2–4]), have been developed to examine the underlying molecular mechanisms responsible for the development and maintenance of neuropathic pain. From these studies, central sensitization (activity-dependent hyperexcitability in some spinal cord neurons) appears to be a critical component in the development of chronic pain states [5], which is driven, at least in part, by abnormal ectopic discharges emanating from damaged primary sensory neurons [6, 7]. While the mechanisms of ectopic discharges are incompletely understood, alterations in the expression and distribution of voltage-gated sodium channels in injured sensory neurons have been implicated as participants in the pathophysiology of chronic pain, providing a mechanistic basis for the clinical use of sodium channel blocker agents as interventions for neuropathic pain [8].

Voltage-gated sodium channels are responsible for action potential electrogenesis in most mammalian neurons, responding to membrane depolarization with transient opening to allow influx of sodium ions. Sodium channels within neurons are composed of a single α-subunit, which forms the voltage-sensing and ion selective pore, and auxiliary β-subunits, which appear to influence channel gating and targeting properties [9–11]. At least ten different mammalian sodium channels have been described [12], of which seven ($Na_V1.1$, $Na_V1.2$, $Na_V1.3$, $Na_V1.6$, $Na_V1.7$, $Na_V1.8$, $Na_V1.9$) are expressed in the nervous system at readily detectable levels during some point of development [13, 14]. All sodium channels share a common motif and considerable homology; however, distinct voltage-dependence, kinetic and pharmacological properties are associated with each of the isoforms (see e.g.,

[15]). Most, if not all, neurons express multiple sodium channel isoforms, and the different repertoires of channels expressed in different types of neurons endow them with unique functional properties.

Primary sensory neurons (dorsal root ganglion (DRG) and trigeminal ganglion neurons) are pseudo-unipolar cells that extend a single process that bifurcates, sending an axon to peripheral targets (e.g., skin, muscle) and also centrally to synapse in the central nervous system (CNS). DRG neurons are a heterogeneous group of neurons, based on morphology (size), sensory modality(ies) transduced, and expression of receptors, neuropeptides and ion channels. Small (<25 μm diameter) DRG neurons are predominantly nociceptive [16–18], and these neurons give rise to unmyelinated (C-type) and thinly-myelinated (Aδ) axons that convey pain information to the CNS in response to noxious stimuli. Nociceptive axons have a relatively high threshold for activation and are generally quiescent unless activated by damaging stimuli [19]. Following injury to peripheral nerves, axons and/or cell bodies of sensory neurons can become hyperexcitable and can give rise to spontaneous action potentials and abnormal high-frequency activity [7, 20–22], which have been suggested to be important contributors to the development of neuropathic pain.

In this chapter, we will review findings that describe alterations in the expression of specific sodium channel isoforms in primary sensory neurons following injury to peripheral nerves. Three experimental models of neuropathic pain will be discussed: nerve transection (neuroma), chronic constriction injury (CCI) and diabetic neuropathy, with the major focus on the expression patterns of sodium channels $Na_V1.3$, $Na_V1.8$ and $Na_V1.9$ in small (<25 μm diameter) DRG neurons following nerve injury. These three sodium channels have received considerable attention due to mounting evidence of their involvement in the pathogenesis of neuropathic pain. We will also review recent observations indicating an important role for $Na_V1.3$ in the development of hyperexcitability in secondary (spinal cord dorsal horn) sensory neurons following peripheral injury (CCI).

Multiple sodium channels in DRG neurons

For nearly 25 years, it has been recognized that DRG neurons express a diversity of sodium currents, which can be discriminated based on their voltage-dependence, kinetic properties and sensitivity to the neurotoxin tetrodotoxin (TTX) [23–26]. In fact, multiple, distinct sodium currents can be recorded within some individual neurons. Studies have demonstrated that most small DRG neurons express fast, TTX-sensitive sodium currents and approximately 85% of these co-express slow, TTX-resistant sodium currents [27]. Recently, specific whole-cell patch clamp recording protocols have been employed that are able to differentiate subpopulations of TTX-sensitive and TTX-resistant currents [27–30], confirming the production of multiple sodium currents by individual DRG neurons.

Reflecting the earlier electrophysiological studies that demonstrated the expression of multiple distinct sodium currents in DRG neurons, it has been shown that normal DRG neurons express transcripts and protein for five sodium channel isoforms (Fig. 1A; [31–33]). Of the seven neuronal sodium channels, $Na_V1.7$, $Na_V1.8$, and $Na_V1.9$ are preferentially expressed in DRG and trigeminal ganglia neurons, and are not found at appreciable levels in the CNS. $Na_V1.7$ is expressed at varying levels in virtually all DRG neurons [31], while channels $Na_V1.8$ and $Na_V1.9$ are predominantly expressed in small DRG neurons [32, 33], with combined electrophysiological subtyping/immunostaining of $Na_V1.8$ [18] and $Na_V1.9$ [17] demonstrating preferential expression in nociceptive neurons. Sodium channels $Na_V1.1$ and $Na_V1.6$, which have widespread expression in the CNS, are also expressed in DRG neurons. Of the channels preferentially expressed in DRG neurons, $Na_V1.8$ and $Na_V1.9$ are distinguished from $Na_V1.1$, $Na_V1.6$ and $Na_V1.7$ by their resistance to block by TTX [32, 33]. $Na_V1.8$ encodes a slowly-inactivating TTX-resistant current (Fig. 1B.B; [32]), while $Na_V1.9$ generates a persistent current with hyperpolarized voltage-dependence of activation and steady state inactivation (Fig. 1B.C; [29]). The TTX-sensitive sodium channels $Na_V1.2$ and $Na_V1.3$ are not expressed above background levels in normal adult DRG neurons. However, as discussed later, sodium $Na_V1.3$ is upregulated in DRG neurons following nerve injury, and along with $Na_V1.8$ and $Na_V1.9$, has been implicated as playing a major, but different, role in neuropathic pain.

Sodium channel expression in DRG neurons during neuroma formation

Transection of peripheral nerve leads to formation of a neuroma, which in its distal 1,000 µm is characterized by a tangle of axonal endbulbs and sprouts, de- and dysmyelinated axons, and extensive disorganized connective tissue [34], and is accompanied by the development of abnormal spontaneous activity (ectopic discharge) in many primary sensory neurons [35, 36]. The aberrant electrical activity can arise at the site of injury [20, 37, 38] or within the DRG cell body [7, 22, 39]. Early studies demonstrated accumulations of sodium channels at the distal tips of transected axons [40–42]. More recent work has identified specific sodium channel isoforms that accumulate within neuromas [43, 44] and that may contribute to hyperexcitability in this region. Transection of the sciatic nerve is also accompanied by alterations in the expression of several sodium channel isotypes in the cell bodies of DRG neurons [33, 45, 46]. Three sodium channel isoforms – $Na_V1.3$, $Na_V1.8$ and $Na_V1.9$ – markedly change their expression patterns following peripheral axotomy, and, due to the unique properties and patterns of distribution of each of these channels, they have received considerable attention for their participation in the development of neuronal hyperexcitability and neuropathic pain following injury to axons.

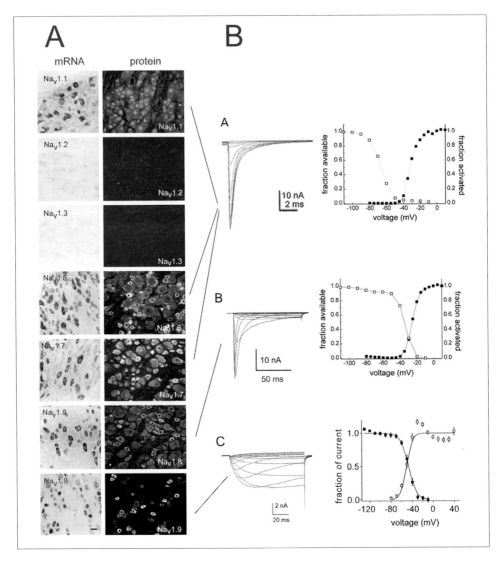

Figure 1

Multiple sodium channels and currents in adult DRG neurons. A. Sodium channel α-subunit mRNAs (left panels) and protein (right panels) visualized by subtype-specific riboprobes and antibodies. Transcripts and protein for five different sodium channels (Na$_V$1.1, Na$_V$1.6, Na$_V$1.7, Na$_V$1.8 and Na$_V$1.9) are present at moderate-to-high levels in DRG neurons. Na$_V$1.2 and Na$_V$1.3 are not detectable in adult DRG neurons. Scale bar, 50 μm. B. Voltage-gated sodium currents recorded by whole-cell patch-clamp in adult DRG neurons. (A) Only fast, TTX-sensitive sodium current (presumably composed of Na$_V$1.1, Na$_V$1.6 and Na$_V$1.7) is observed in a muscle afferent DRG neuron, which exhibits little overlap between activation (filled cir-

The TTX-sensitive sodium channel $Na_V1.3$ is virtually undetectable in normal adult DRG and trigeminal neurons. However, early studies demonstrated a significant upregulation of $Na_V1.3$ transcripts in DRG neurons following peripheral axotomy (Fig. 2A; [45, 46]). More recently, the upregulation of $Na_V1.3$ mRNA has been shown to be accompanied with increased expression of $Na_V1.3$ protein in DRG neurons (Fig. 2B; [43]). In parallel with $Na_V1.3$ mRNA and protein upregulation following injury to DRG neurons, there is an emergence of a rapidly-repriming (i.e., recovers rapidly from inactivation) TTX-sensitive sodium current (Fig. 2C a and b; [27]). The concurrent upregulation of $Na_V1.3$ and the emergence of the rapidly-repriming current led to the suggestion that $Na_V1.3$ is responsible for the rapidly-repriming current [27]. Additional support for this suggestion is provided by the appearance of a rapidly-repriming current in HEK 293 cells and DRG neurons when they are transfected with a $Na_V1.3$ construct [47]. It has been suggested [27] that the rapid recovery from inactivation displayed by this channel should support sustained high frequency firing [48], which could contribute markedly to neuronal hyperexcitability.

Upregulation of $Na_V1.3$ in transected DRG neurons appears to be a specific response of this population of neurons to peripheral nerve injury and does not mirror an upregulation of this channel in other classes of neurons after axotomy. Axotomy within the spinal cord of primary motor neurons is not accompanied by upregulation of $Na_V1.3$ in these neurons [49], and transection of the sciatic nerve does not result in enhanced $Na_V1.3$ expression in ventral horn spinal motoneurons (Black and Waxman, unpublished observations). Moreover, transection of the central projections of DRG neurons (dorsal rhizotomy) is not accompanied by $Na_V1.3$ upregulation [43].

Importantly, accompanying the upregulation of $Na_V1.3$ within the cell bodies of peripherally-axotomized DRG neurons, $Na_V1.3$ has been shown to accumulate within the neuroma of transected sciatic nerve (Fig. 2D; [43]). $Na_V1.3$ immunoreactivity is localized to the distal region of the transected nerve, with only background levels of immunofluorescence greater than 500–1,000 µm proximal to this region. The specific aggregation of $Na_V1.3$ within the neuroma targets this channel, due to its rapidly repriming kinetics, to play an active role in the generation of ectopic discharges which are known to emanate from this region [37, 50, 51].

---

cles) and steady-state inactivation (unfilled circles). (B) A small DRG neuron displays only slow, TTX-resistant sodium current ($Na_V1.8$); activation and steady-state inactivation curves are depolarized compared to fast, TTX-sensitive current. (C) Persistent, TTX-resistant sodium current ($Na_V1.9$) recorded from a small DRG neuron from $Na_V1.8$-null mouse. Activation (unfilled circles) and steady-state inactivation (filled circles) show significant overlap (window currents). (Modified and reproduced with permission from [29, 31, 54, 90, 91])

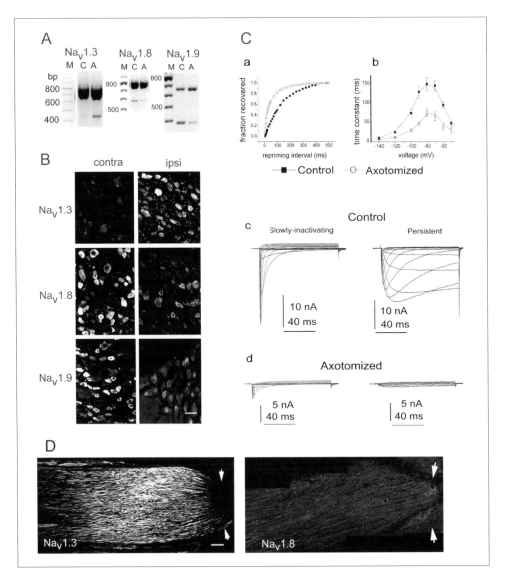

Figure 2

Alterations in expression of Na$_V$1.3, Na$_V$1.8 and Na$_V$1.9 in DRG neurons following peripheral transection of sciatic nerve. A. RT-PCR analyses of control (C) and peripherally-axotomized (A) DRG demonstrates upregulation of Na$_V$1.3 and downregulation of Na$_V$1.8 and Na$_V$1.9 at 7–12 days following axotomy. B. Contralateral (contra) and ipsilateral (ipsi) DRG reacted with isoform-specific antibodies for Na$_V$1.3, Na$_V$1.8 and Na$_V$1.9 display an upregulation of Na$_V$1.3 signal and a downregulation of immunofluorescent signal for Na$_V$1.8 and Na$_V$1.9 within DRG neurons. Scale bar, 50 µm. C. Whole-cell patch-clamp recordings of

In contrast to the upregulation of the TTX-sensitive channel $Na_V1.3$ following axotomy, the two TTX-resistant channels within DRG neurons, $Na_V1.8$ and $Na_V1.9$, exhibit significant downregulation of their transcripts (Fig. 2A; [33, 46, 52]) and protein (Fig. 2B; [54]) following peripheral axotomy. The reduction of transcripts and protein for the TTX-resistant channels $Na_V1.8$ and $Na_V1.9$ in axotomized DRG neurons is consistent with early studies demonstrating a reduction in *total* TTX-resistant current in these neurons following peripheral axotomy [27, 55]. The development of specific whole-cell patch-clamp protocols has made it possible to separate the slowly-inactivating current attributable to $Na_V1.8$ from the persistent current of $Na_V1.9$ [29], and to demonstrate that both the slowly-inactivating $Na_V1.8$ and persistent $Na_V1.9$ TTX-resistant currents are attenuated in axotomized DRG neurons [54].

As might be anticipated from the downregulation of $Na_V1.8$ and $Na_V1.9$ mRNA and protein in DRG neurons following peripheral axotomy, these channels do not accumulate within the neuroma at 9–14 days following sciatic nerve transection (e.g., Fig. 2D). In this respect, it is not entirely clear what role(s) these channels may play, if any, in the generation of ectopic discharges, and in the development and maintenance of neuropathic pain. While it has been established that ectopic activity may occur in both C-type [56–59] and A-fibers [60, 61], application of nanomolar concentrations of TTX is reported to silence most ectopic discharges [62, 63], which would not inhibit $Na_V1.8$ or $Na_V1.9$ activity, since these channels have K_Ds of ~40–60 µM [29, 32]. Interestingly, spontaneous ectopic discharge is extremely rare (0.4%) in neuromas of $Na_V1.8$-null mice [16] at 22 days post-transection compared to WT mice (18%; [64]), suggesting an involvement of $Na_V1.8$ in abnormal firing in these fibers. Moreover, intrathecal administration of $Na_V1.8$ antisense oligodeoxynucleotides in L5/L6 spinal nerve ligated rats is reported to ameliorate neuropathic pain in these animals [65], although it is unclear whether this treatment affected ectopic discharge.

Like $Na_V1.8$, $Na_V1.9$ also does not accumulate in neuromas at 9–14 days (Black and Waxman, unpublished observations). $Na_V1.9$ has been shown to be expressed selectively in most, but not all, C-type and A-fiber nociceptive-type DRG neurons

---

control and axotomized small DRG neurons. (a) The time course for recovery from inactivation at −80 mV is faster in axotomized (open circles) than control (filled circles) neurons. (b) Time constants for recovery from activation plotted as function of voltage. Time constants for axotomized (open circles) are smaller than for control (filled circles) neurons. (c) (d) Slowly inactivating ($Na_V1.8$) and persistent ($Na_V1.9$) TTX resistant currents are reduced in small DRG neurons following peripheral axotomy. D. $Na_V1.3$ immunostaining is present within the neuroma immediately proximal to the sciatic nerve ligature (arrowheads) and transection; $Na_V1.8$ does not accumulate within the neuroma. Scale bar, 100 µm. (Modified and reproduced with permission from [33, 43, 46, 54])

[17], and is localized within C-type fibers and nerve endings [66–68]. The hyperpolarized voltage-dependence of activation (threshold, ~–70 mV; midpoint, –41 mV) and steady state-inactivation (midpoint, –44 mV), and the substantial overlap of the activation and steady-state inactivation curves for $Na_V1.9$ (Fig. 1B.D.), are predicted to give rise to a persistent current (window current) that is active near the resting potential [29]; computer simulations of DRG neurons that incorporate TTX-sensitive and TTX-resistant currents suggest that $Na_V1.9$ depolarizes the membrane by 10–20 mV and enhances the response to depolarizing inputs that are subthreshold for spike generation [69]. Baker et al. [70] used current-clamp recording to show that the $Na_V1.9$ current, which is upregulated by GTP, increases the excitability of small DRG neurons, with upregulation of the current reducing threshold and leading to the generation of spontaneous firing in neurons that had been silent. On the other hand, it has also been suggested [27] that downregulation of $Na_V1.9$ following axotomy, and the subsequent loss of its depolarizing effect, can hyperpolarize the resting membrane potential and thereby release resting inactivation of TTX-sensitive sodium channels, contributing to hyperexcitability. In conjunction with an upregulation and targeting of $Na_V1.3$ in the neuroma, the downregulation of $Na_V1.9$ after nerve injury may therefore enhance susceptibility to ectopic discharge.

Sodium channel expression in DRG neurons in the CCI model of peripheral injury

The rodent chronic constriction injury (CCI) is a well-established model that has been utilized to examine the mechanisms underlying neuropathic pain [2]. CCI results in Wallerian degeneration of a substantial number of, but not all, axons distal to the loose ligatures, with greater than 80% loss of myelinated fibers and 60–80% loss of unmyelinated fibers [71]. Proximal to the loose ligatures, the proximal stumps of degenerating axons intermingle with spared axons [71, 72], leading to injured and uninjured neurons residing in L4 and L5 DRG. Behaviorally, CCI is associated with signs of spontaneous pain and mechanical hyperalgesia [2]; abnormal spontaneous activity has been recorded *in vivo* and *in vitro* in some DRG neurons following CCI [22, 57, 73, 74]. Current evidence strongly suggests that alterations in sodium channel expression contribute to the spontaneous activity observed in CCI neurons, which may play a major role in the development of ongoing and stimulus-driven neuropathic pain [75]. Moreover, recent evidence indicates that CCI also affects the excitability properties of second order sensory neurons within the spinal cord entry zone, which contributes to the hyperalgesia and allodynia following CCI [76].

In comparison to sciatic nerve transection, alterations in the expression of sodium channels $Na_V1.3$, $Na_V1.8$ and $Na_V1.9$ in DRG neurons are similar, but less extensive, following CCI [53, 75, 77]. As demonstrated by RT-PCR analysis of L4/5

DRG from control and CCI rats 14 days post-surgery, $Na_V1.3$ is upregulated and $Na_V1.8$ and $Na_V1.9$ are downregulated following this injury (Fig. 3A). Consistent with these results, *in situ* hybridization studies performed on DRG neurons cultured from control and CCI rats demonstrate significantly enhanced signal for $Na_V1.3$ and attenuated signals for $Na_V1.8$ and $Na_V1.9$ (Fig. 3B). In parallel with an upregulation of $Na_V1.3$, there is an emergence of rapidly-repriming current in CCI neurons (Fig. 3C), similar to that occurring after axotomy [27, 43]. The attenuation in the levels of $Na_V1.8$ and $Na_V1.9$ transcripts in CCI neurons is accompanied by a significant reduction in total TTX-resistant current. These changes in sodium channel expression would be expected to alter the firing properties of injured DRG neurons, similar to the changes occurring following axotomy. At this point, however, no data are available to indicate whether $Na_V1.3$ accumulates in injured axons proximal to the loose ligatures.

Sodium channel expression in spinal cord dorsal horn neurons following peripheral nerve injury

It has been known for some time that, in response to peripheral injury, dorsal horn neurons undergo reactive changes that make them hyperresponsive and display abnormal firing properties [78–81]. An involvement of sodium channels in the changes in electrogenesis in these dorsal horn neurons has been suggested, but has not been elucidated. Recently, Hains et al. [76] have provided strong evidence that $Na_V1.3$ plays a major role in the hyperexcitability of dorsal horn neurons following peripheral injury. As shown in Figure 4, $Na_V1.3$ mRNA is not detectable within laminas I–V of the dorsal horn of control rats; however, 10 days following CCI, significant $Na_V1.3$ hybridization signal is present with small (5–10 μm) neurons in laminas I–II and in larger (20–40 μm) neurons in laminas II–V. RT-PCR analysis confirmed an upregulation of $Na_V1.3$ in dorsal horn ipsilateral to CCI compared to contralateral CCI or control dorsal horn (Fig. 4D). Co-localization studies with antibodies against $Na_V1.3$ and NK1R, a marker for nociceptive neurons, demonstrated substantial co-localization within the dorsal horn neurons. At this same post-surgical time, the CCI rats exhibit allodynia and hyperalgesia, and extracellular unit recordings in the dorsal horn demonstrate abnormal spontaneous firing and increased evoked activities in response to all peripheral stimuli (brush, pressure, pinch).

To examine the role of $Na_V1.3$ in dorsal horn hyperresponsiveness and behavioral changes, targeted oligodeoxynucleotide (ODN) knockdown of $Na_V1.3$ was performed *via* intrathecal administration of $Na_V1.3$ anti-sense (AS) and mismatch (MM) for 4 days beginning 11 days following CCI. While AS administration was not effective in knocking down $Na_V1.3$ mRNA expression in ipsilateral CCI DRG neurons, $Na_V1.3$ AS significantly decreased $Na_V1.3$ mRNA signal in ipsilateral dor-

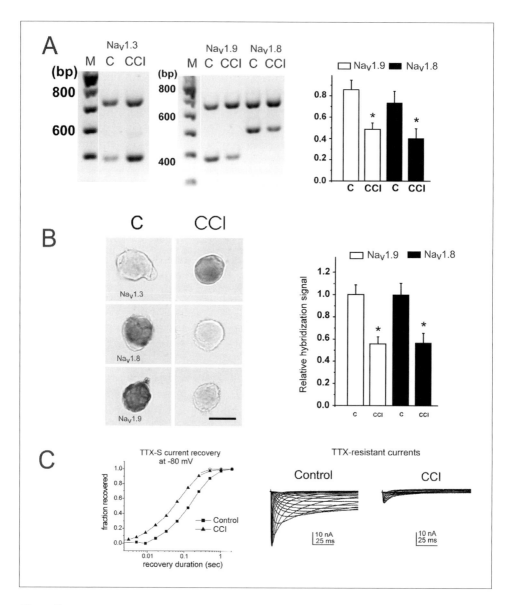

Figure 3

Chronic constriction injury (CCI) of sciatic nerve upregulates Na$_V$1.3 and downregulates Na$_V$1.8 and Na$_V$1.9 in DRG neurons. A. Quantification of RT-PCR products from control (C) and CCI DRG 14 days post surgery demonstrates a significant downregulation of Na$_V$1.8 and Na$_V$1.9 mRNA (Na$_V$1.8: 512 bp; Na$_V$1.9: 392 bp). Na$_V$1.3 transcripts are upregulated in CCI neurons (Na$_V$1.3: 412 bp). B. In situ hybridization of small neurons from control and CCI DRG shows that hybridization signals for Na$_V$1.8 and Na$_V$1.9 are significantly reduced in CCI

sal horn neurons (Fig. 5B and C). MM administration had no effect on the expression of $Na_V1.3$ in dorsal horn neurons (Fig. 5A). In parallel with the attenuation of $Na_V1.3$ signal in dorsal horn neurons following AS treatment, unit recordings in the dorsal horn demonstrated a reduction in spontaneous activity and decreased hyperresponsiveness following AS, but not MM, administration (Fig. 5D). Behavioral testing demonstrated that AS administration for 4 days starting on day 11 decreased mechanical allodynia and increased the withdrawal threshold (Fig. 5E). Importantly, following cessation of AS treatment, allodynia and thermal hyperalgesia returned to pre-AS treatment values. These observations provide strong evidence that the upregulated expression of $Na_V1.3$ within dorsal horn sensory neurons contributes to hyperresponsiveness of these neurons and to resultant allodynia and hyperalgesia.

Sodium channel expression in experimental painful diabetic neuropathy

Peripheral neuropathies often accompany disease states, including diabetes mellitus, herpes zoster and HIV, and are a major cause of pain syndromes in these patients. While the molecular mechanisms underlying the neuropathic pain are largely unknown, recent work with a model of peripheral neuropathy, the rodent streptozotocin-induced diabetic neuropathy, has provided some insight into processes that may contribute to the painful condition. In these studies, alterations of sodium channel expression have been linked to the development of painful neuropathy [53, 82, 83], and may provide a molecular mechanism underlying this condition.

Craner et al. [82] studied the rodent streptozotocin (STZ)-induced diabetes, a well-established model for the study of diabetic neuropathy [84–87]. In this model, injection of STZ results in a significant increase in blood glucose levels after 4 days (479 ± 27 mg/dl diabetic compared to 126 ± 3 mg/dl control), which is maintained for at least 8 weeks (513 ± 29 mg/dl) [82]. The withdrawal threshold falls by 3 weeks following STZ injection (8.6 ± 1.7 gm diabetic *versus* 17.3 ± 2.4 gm control) and generally reaches the threshold for tactile allodynia (≥ 4.0 gm) approximately 6 weeks following injection. A qualitative examination of the transcript levels of $Na_V1.1$, $Na_V1.3$, $Na_V1.6$, $Na_V1.7$, $Na_V1.8$ and $Na_V1.9$ by *in situ* hybridization at 1-week and 8-weeks following the onset of allodynia indicated no change in expres-

neurons, while hybridization signal for Na$_V$1.3 is upregulated. Scale bar, 20 μm. C. Recovery from inactivation is accelerated after CCI; repriming kinetics of TTX-sensitive current is significantly increased in CCI neurons compared to control neurons. Patch-clamp recording demonstrates a significant reduction in TTX-resistant currents in CCI neurons compared to control neurons. (Modified and reproduced with permission from [75])

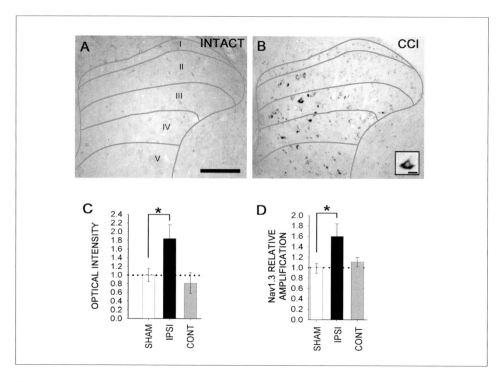

Figure 4
Chronic constriction injury (CCI) of sciatic nerve upregulates Na$_V$1.3 in second order senso-
ry neurons in dorsal horn gray matter. A. In situ hybridization for Na$_V$1.3 transcripts reveals
no hybridization signal within dorsal horn gray matter in sham-operated (intact) rats. B. Ten
days following CCI, Na$_V$1.3 signal is present within cells in the superficial and deep ipsilat-
eral dorsal horn. Inset, labeled cell exhibits multipolar neuronal profile and cytoplasmic
staining. Scale bar, A., B. 300 μm; inset, 10 μm. C. Quantification of hybridization signal
demonstrates a significant increase in Na$_V$1.3 label in CCI ipsilateral (ipsi) neurons compared
to CCI contralateral (cont) or sham-operated (sham) neurons. D. Quantitative RT-PCR
demonstrates a significant increase in Na$_V$1.3 amplification signal in CCI ipsilateral (ipsi)
dorsal horn compared to CCI contralateral (cont) or sham-operated (sham) dorsal horn.
(Modified and reproduced with permission from [76])

Figure 5 (see next page)
Intrathecal administration of Na$_V$1.3 antisense (AS) oligodeoxynucleotide (ODN) attenuates
Na$_V$1.3 signal and neuropathic pain following CCI. A.,B. Four days following administration
of Na$_V$1.3 mismatch (MM) or AS ODNs, beginning 11 days after CCI, in situ hybridization
signal was unchanged in MM rats compared to CCI (no ODN) rats, but was markedly
reduced in AS rats. Scale bar, 300 μm. C. Quantitative RT-PCR demonstrates a significant

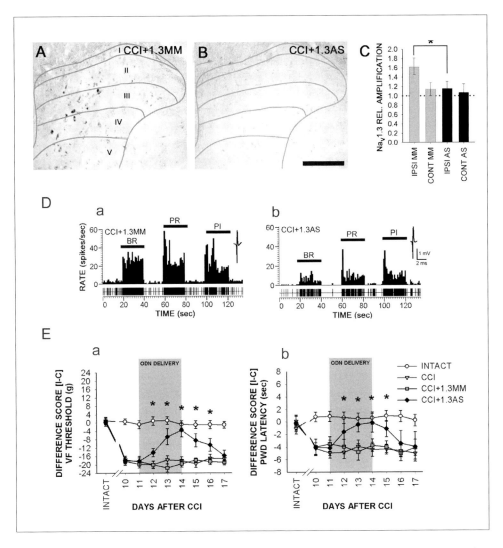

reduction in $Na_V1.3$ signal in ipsilateral dorsal horn in rats receiving $Na_V1.3$ AS compared to rats receiving $Na_V1.3$ MM. D. After 4 days of intrathecal administration, beginning 11 days following CCI, $Na_V1.3$ MM had no effect on spontaneous or evoked activity (a), but $Na_V1.3$ AS attenuated both spontaneous and evoked activities (b); spontaneous and evoked discharge rates show significantly decreased evoked activity to all peripheral stimuli following CCI in $Na_V1.3$ AS rats. E. Pain-related behaviors after CCI and $Na_V1.3$ AS or MM administration. Mechanical allodynia (a) and thermal hyperalgesia (b) in CCI rats are attenuated by $Na_V1.3$ AS but not MM administration (gray shaded area). Cessation of $Na_V1.3$ AS administration is accompanied by increase in mechanical allodynia and thermal hyperalgesia. (Modified and reproduced with permission from [76])

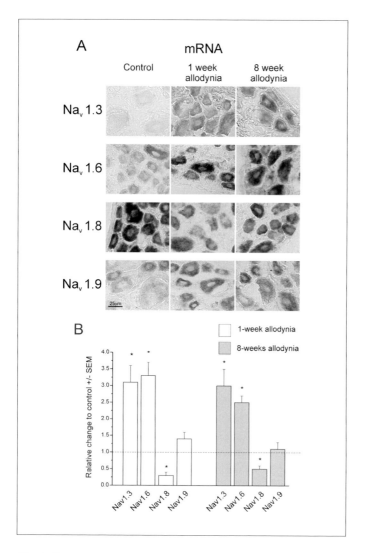

Figure 6
Alterations in expression of sodium channels within DRG in painful diabetic neuropathy. A.
In situ hybridization demonstrates that transcripts for Na_V1.3 and Na_V1.6 are significantly
upregulated in small DRG neurons at 1- and 8-weeks post allodynia in diabetic neuropathy
rats compared to control rats. Na_V1.8 mRNA is significantly reduced in small DRG neurons
at 1- and 8-weeks post allodynia, while the signal for Na_V1.9 mRNA is not substantially
changed at these times. There is an increase in expression of Na_V1.9 within large DRG neu-
rons. Scale bar, 25 μm. B. Histogram showing relative changes in immunostaining intensity
for Na_V1.3, Na_V1.6, Na_V1.8 and Na_V1.9 within small DRG neurons in STZ-diabetic neu-
ropathy. (Modified and reproduced with permission from [82])

sion for $Na_V1.1$ and $Na_V1.7$ [82]. However, there was a significant upregulation of $Na_V1.3$ and $Na_V1.6$ and a downregulation $Na_V1.8$ in small < 25 µm) and large (> 25 µm) DRG neurons, and a non-significant slight increase in $Na_V1.9$ in small (< 25 µm diameter) DRG neurons, which are largely nociceptive, at 1- and 8-weeks following the onset of allodynia (Fig. 6). Protein levels generally mirrored transcript levels in the diabetic neurons. Downregulation of $Na_V1.8$ in STZ-induced diabetic neurons has been observed in several reports [53, 82, 83], and is similar to what occurs following axotomy [46]. Craner et al. [82] also observed a significant upregulation of $Na_V1.9$ mRNA and protein in large (> 25 µm) DRG neurons at 1- and 8-weeks following the onset on allodynia.

The changes in channel expression in the diabetic DRG neurons may contribute to hyperexcitability of these neurons. $Na_V1.3$, as discussed previously, produces a rapidly-repriming sodium current, which would support sustained high frequency firing. Moreover, $Na_V1.6$ produces both a fast transient current and a smaller persistent current that is generated closer to the resting potential [88]. This persistent current can contribute to burst activities, and the upregulation of $Na_V1.6$, as well as the upregulation of $Na_V1.9$, would be anticipated to contribute to hyperexcitability of the neurons.

Perspectives

Current evidence clearly implicates voltage-gated sodium channels as playing a major role in the pathogenesis of chronic pain following injury to peripheral nerves and peripheral neuropathies. Alterations in the expression and targeting of specific sodium channels within primary sensory neurons appear to predispose these neurons to abnormal firing properties, leading to the development of neuropathic pain. The current evidence suggests that the selective knockout or rescue of specific sodium channels may attenuate neuropathic pain without adversely affecting vital sodium channel functions. Indeed, glial cell line-derived neurotrophic factor (GDNF) has been shown to ameliorate sensory abnormalities that develop following injury to peripheral nerves, which may reflect the rescue of $Na_V1.8$ and/or $Na_V1.9$ expression and the suppression of $Na_V1.3$ expression in DRG neurons [60, 89]. It is anticipated that, as more is understood of the mechanisms governing their expression and the molecular architecture responsible for their differing biophysical properties, sodium channels will emerge as critical therapeutic targets for the management of neuropathic pain.

Acknowledgements
Research in the authors' laboratory has been supported in part by grants from the Rehabilitation Research Service and the Medical Research Service, Department of

Veterans Affairs, and from the National Multiple Sclerosis Society. The Center for Neuroscience and Regeneration Research is a Collaboration of the Paralyzed Veterans of America and The United Spinal Association with Yale University.

References

1   Arner S, Meyerson BA (1988) Lack of analgesic effect of opioids on neuropathic and idiopathic forms of pain. *Pain* 33: 11–23

2   Bennett GJ, Xie YK (1988) A peripheral mononeuropathy in rat that produces disorders of pain sensation like those seen in man. *Pain* 33: 87–107

3   Seltzer Z, Dubner R, Shir Y (1990) A novel behavioral model of neuropathic pain disorders produced in rats by partial sciatic nerve injury. *Pain* 43: 205–218

4   Kim SH, Chung JM (1992) An experimental model for peripheral neuropathy produced by segmental spinal nerve ligation in the rat. *Pain* 50: 355–362

5   Dubner R, Ruda MA (1992) Activity dependent neuronal plasticity following tissue injury and inflammation. *Trends Neurosci* 15(3): 96–103

6   Choi Y, Yoon YW, Na HS, Kim SH, Chung JM (1994) Behavioral signs of ongoing pain and cold alloynia in a rat model of neuropathic pain. *Pain* 59: 369–376

7   Devor M, Janig W, Michaelis M (1994) Modulation of activity in dorsal root ganglion neurons by sympathetic activation in nerve-injured rats. *J Neurophysiol* 71: 38–47

8   Dickenson AH, Matthews MA, Suzuki R (2002) Neurobiology of neuropathic pain: mode of action of anticonvulsants. *Eur J Pain* 6 (Suppl A): 51–60

9   Qu Y, Curtis R, Lawson D, Gilbride K, Ge P, DiStefano PS, Silos-Santiago I, Catterall WA, Scheuer T (2001) Differential modulation of sodium channel gating and persistent sodium currents by the β1, β2, and β3 subunits. *Mol Cell Neurosci* 18: 570–580

10 Chen C, Bharucha V, Chen Y, Westenbroek RE, Brown A, Malhotra JD, Jone D, Avery C, Gillespie III, PJ, Kazen-Gillespie KA et al (2002) Reduced sodium channel density, altered voltage dependence of inactivation, and increased susceptibility to seizures in mice lacking sodium channel β2-subunits. *Proc Natl Acad Sci USA* 99: 17072–17077

11 Yu FH, Westenbroek RE, Silos-Santiago I, McCormick KA, Lawson D, Ge P, Ferriera H, Lilly J, DiStefano PS, Catterall WA et al (2003) Sodium channel beta4, a new disulfide-linked auxiliary subunit with similarity to beta2. *J Neurosci* 23: 7577–7585

12 Goldin AL, Barchi RL, Caldwell JH, Hofmann F, Howe JR, Hunter JC, Kallen RG, Mandel G, Meisler MH, Netter YB et al (2000) Nomenclature of voltage-gated sodium channels. *Neuron* 28: 365–368

13 Felts PA, Yokoyama S, Dib-Hajj S, Black JA, Waxman SG (1997) Sodium channel α-subunit mRNAs I, II, III, NaG, Na6, hNE (PN1): different expression patterns in developing rat nervous system. *Mol Brain Res* 45: 71–82

14 Amaya F, Decosterd I, Samad TA, Plumpton C, Tate S, Mannion RJ, Costigan M, Woolf CJ (2000) Diversity of expression of the sensory neuron-specific TTX-resistant voltage-gated sodium ion channels SNS and SNS2. *Mol Cell Neurosci* 15: 331–342

15 Yu FH, Catterall WA (2003) Overview of the voltage-gated sodium channel family. *Genome Biol* 4: 207–207.7

16 Akopian AN, Souslova V, England S, Okuse K, Ogata N, Ure J, Smith A, Kerr BJ, McMahon SB, Boyce S et al (1999) The tetrodotoxin-resistant sodium channel SNS has a specialized function in pain pathways. *Nat Neurosci* 2: 541–548

17 Fang X, Djouhri L, Black JA, Dib-Hajj SD, Waxman SG, Lawson SN (2002) The presence and role of the tetrodotoxin-resistant sodium channel $Na_V1.9$ (NaN) in nociceptive primary afferent neurons. *J Neurosci* 22: 7425–7433

18 Djouhri L, Fang X, Okuse K, Wood JN, Berry CM, Lawson SN (2003) The TTX-resistant sodium channel $Na_V1.8$ (SNS/PN3): expression and correlation with membrane properties in rat nociceptive primary afferent neurons. *J Physiol* 550: 739–752

19 Lawson SN (2002) Phenotype and function of somatic primary afferent nociceptive neurones with C-, Aδ and β fibres. *Exp Physiol* 87: 239–244

20 Wall PD, Gutnick M (1974) Ongoing activity in peripheral nerves: the physiology and pharmacology of impulses originating from a neuroma. *Exp Neurol* 43: 580–593

21 Wall PD, Devor M (1983) Sensory afferent impulses originate from dorsal root ganglia as well as from the periphery in normal and nerve injured rats. *Pain* 17: 321–339

22 Kajander KC, Wakisaka S, Bennett GJ (1992) Spontaneous discharge originates in the dorsal root ganglion at the onset of a painful peripheral neuropathy in the rat. *Neurosci Lett* 138: 225–228

23 Kostyuk PG, Veselovsky NS, Tsyandryenko AY (1981) Ionic currents in the somatic membrane of rat dorsal root ganglion neurons. I. sodium currents. *Neuroscience* 12: 2423–2430

24 Roy ML, Narahashi T (1992) Differential properties of tetrodotoxin-sensitive and tetrodotoxin-resistant sodium channels in rat dorsal root ganglion neurons. *J Neurosci* 12: 2104–2111

25 Caffrey JM, Eng DL, Black JA, Waxman SG, Kocsis JD (1992) Three types of sodium channels in adult rat dorsal root ganglion neurons. *Brain Res* 592: 283–297

26 Elliott AA, Elliott JR (1993) Characterization of TTX-sensitive and TTX-resistant sodium currents in small cells from adult rat dorsal root ganglia. *J Physiol (Lond)* 463: 39–56

27 Cummins TR, Waxman SG (1997) Downregulation of tetrodotoxin-resistant sodium current in small spinal sensory neurons after nerve injury. *J Neurosci* 17: 3503–3514

28 Cummins TR, Howe JR, Waxman SG (1998) Slow closed-state inactivation: a novel mechanism underlying ramp currents in cells expressing the hNE/PN1 sodium channel. *J Neurosci* 18: 9607–9619

29 Cummins TR, Dib-Hajj SD, Black JA, Akopian AN, Wood JN, Waxman SG (1999) A novel persistent tetrodotoxin-resistant sodium current in small primary sensory neurons. *J Neurosci* 19: RC43–48

30 Smith RD, Goldin AL (1998) Functional analysis of the rat I sodium channel in xenopus oocytes. *J Neurosci* 18: 811–829

31 Black JA, Dib-Hajj S, McNabola K, Jeste S, Rizzo MA, Kocsis JD, Waxman SG (1996) Spinal sensory neurons express multiple sodium channel α-subunit mRNAs. *Mol Brain Res* 43: 117–132

32   Akopian AN, Sivilotti L, Wood JN (1996) A tetrodotoxin-resistant voltage-gated sodium channel expressed by sensory neurons. *Nature* 379: 257–262

33   Dib-Hajj SD, Tyrrell L, Black JA, Waxman SG (1998) NaN, a novel voltage-gated Na channel, is expressed preferentially in peripheral sensory neurons and down-regulated after axotomy. *Proc Natl Acad Sci USA* 95: 8963–8969

34   Fried K, Govrin-Lippmann R, Rosenthal F, Ellisman MH, Devor M (1991) Ultrastructure of afferent axon endings in a neuroma. *J Neurocytol* 20: 682–701

35   Grovrin-Lippmann R, Devor M (1978) Ongoing activity in severed nerves: source and variation with time. *Brain Res* 159(2): 406–410

36   Ochoa JL, Torebjork HE (1980) Paraesthesiae from ectopic impulse generation in human sensory nerves. *Brain* 103(4): 835–853

37   Burchiel KJ (1984) Spontaneous impulse generation in normal and denervated dorsal root ganglia: sensitivity to alpha-adrenergic stimulation and hypoxia. *Exp Neurol* 85(2): 257–272

38   Devor M, Janig W (1984) Activation of myelinated afferents ending in a neuroma by stimulation of the sympathetic supply in the rat. *Neurosci Lett* 24(1): 43–47

39   Liu X, Cjung K, Chung JM (1999) Ectopic discharges and adrenergic sensitivity of sensory neurons after spinal nerve injury. *Brain Res* 849: 244–247

40   Devor M, Keller CH, Deerinck TJ, Ellisman MH (1989) Na$^+$ channel accumulation on axolemma of afferent endings in nerve end neuromas in Apteronotus. *Neurosci Lett* 102: 149–154

41   England JD, Gamboni F, Ferguson MA, Levinson SR (1994) Sodium channels accumulate at the tips of injured axons. *Muscle & Nerve* 17: 593–598

42   England JD, Happel LT, Kline DG, Gamboni F, Thouron CL, Lui ZP, Levinson SR (1996) Sodium channel accumulation in humans with painful neuromas. *Neurology* 47: 272–276

43   Black JA, Cummins TR, Plumpton C, Chen Y, Clare J and Waxman SG (1999) Upregulation of a previously silent sodium channel in axotomized DRG neurons. *J Neurophysiol* 82: 2776–2785

44   Kretschmer T, England JD, Happel LT, Liu AP, Thouron CL, Nguyen DH, Beuerman RW, Kline DG (2002) Ankyrin G and voltage-gated sodium channels colocalize in human neuroma – key proteins of membrane remodeling after axonal injury. *Neurosci Lett* 323: 151–155

45   Waxman SG, Kocsis JK, Black JA (1994) Type III sodium channel mRNA is expressed in embryonic but not adult spinal sensory neurons, and is reexpressed following axotomy. *J Neurophysiol* 72: 466–471

46   Dib-Hajj S, Black JA, Felts P, Waxman SG (1996) Down-regulation of transcripts for Na channel α-SNS in spinal sensory neurons following axotomy. *Proc Natl Acad Sci USA* 93: 14950–14954

47   Cummins TR, Alieco F, Renganathan M, Herzog RI, Dib-Hajj SD, Waxman SG (2001) Na$_V$1.3 sodium channels: rapid-repriming and slow closed-state inactivation display quantitative differences after expression in mammalian cell line and in spinal sensory neurons. *J Neurosci* 21: 5952–5961

48 Chahine M, George AL Jr, Zhou M, Ji S, Sun W, Barchi RL, Horn R (1994) Sodium channel mutations in paramyotonia congenita uncouple inactivation from activation. *Neuron* 12(2): 281–294

49 Hains BC, Black JA, Waxman SG (2002) Primary motor neurons fail to un-regulate voltage-gated sodium channel Na$_V$1.3/brain type III following axotomy resulting from spinal cord injury. *J Neurosci Res* 70: 546–552

50 Scadding JW (1981) Development of ongoing activity, mechanosensitivity, and adrenaline sensitivity in severed peripheral nerve axons. *Exp Neurol* 173(2): 345–364

51 Matzner O, Devor M (1994) Hyperexcitability at sites of nerve injury depends on voltage-sensitive Na$^+$ channels. *J Neurophysiol* 72(1): 349–359

52 Tate S, Benn S, Hick C, Trezise D, John V, Mannion RJ, Costigan M, Plumpton C, Grose D, Gladwell Z et al (1998) Two sodium channels contribute to the TTX-R sodium current in primary sensory neurons. *Nat Neurosci* 1: 653–655

53 Okuse K, Chaplan SR, McMahon SB, Luo ZD, Calcutt NA, Scott BP, Akopian AN, Wood JN (1997) Regulation of expression of the sensory neurons-specific sodium channel SNS in inflammatory and neuropathic pain. *Mol Cell Neurosci* 10: 196–207

54 Sleeper AA, Cummins TR, Hormuzdiar W, Tyrrell L, Dib-Hajj SD, Waxman SG, Black JA (2000) Changes in expression of two tetrodotoxin-resistant sodium channels and their currents in dorsal root ganglion neurons following sciatic nerve injury, but not rhizotomy. *J Neurosci* 20: 7279–7289

55 Rizzo MA, Kocsis JD, Waxman SG (1995) Selective loss of slow and enhancement of fast Na$^+$ currents in cutaneous afferent dorsal root ganglion neurones following axotomy. *Neurobiol Dis* 2: 87–96

56 Zhang J-M, Donnelly DF, Song X-J, LaMotte RH (1997) Axotomy increases the excitability of dorsal root ganglion cells with unmyelinated axons. *J Neurophysiol* 78: 2790–2794

57 Kajander KJ, Bennett GJ (1992) Onset of a painful peripheral neuropathy in rat: a partial and differential deafferentation and spontaneous discharge in A beta and A delta primary afferent neurons. *J Neurophysiol* 68(3): 734–744

58 Study RE, Kral MG (1996) Spontaneous action potential activity in isolated dorsal root ganglion neurons from rats with a painful neuropathy. *Pain* 65: 235–242

59 Wu G, Ringkamp M, Hartke TV, Murinson BB, Campbell JN, Griffin JW, Meyer RA (2001) Early onset of spontaneous activity in uninjured C-fiber nociceptors after injury to neighboring nerve fibers. *J Neurosci* 21: RC140: 1–5

60 Boucher TJ, Okuse K, Bennett DL, Munson JB, Wood JN, Mcmahon SB (2000) Potent analgesic effects of GDNF in neuropathic pain states. *Science* 290: 124–127

61 Liu C-N, Wall PD, Den-Dor E, Michaelis M, Amir R, Devor M (2000) Tactile allodynia in the absence of C-fiber activation; altered firing properties of DRG neurons following spinal nerve injury. *Pain* 85: 503–521

62 Omana-Zapata I, Khabbaz MA, Hunter JC, Clarke DE, Bley KR (1997) Tetrodotoxin inhibits neuropathic ectopic activity in neuromas, dorsal root ganglia and dorsal horn neurons. *Pain* 72: 41–49

63 Liu X, Zhou J-L, Chung K, Chung JM (2001) Ion channels associated with the ectopic discharges generated after segmented spinal nerve injury in the rat. *Brain Res* 900: 119–127

64 Roza C, Laird JMA, Souslova V, Wood JN, Cervero F (2003) The tetrodotoxin-resistant Na$^+$ channel Na$_V$1.8 is essential for the expression of spontaneous activity in damaged sensory axons of mice. *J Physiol (Lond)* 550: 921–926

65 Lai J, Gold MS, Kim C-S, Bian D, Ossipov H, Hunter JC, Porreca F (2002) Inhibition of neuropathic pain by decreased expression of the tetrodotoxin-resistant sodium channel, Na$_V$1.8. *Pain* 95: 143–152

66 Fjell J, Hjelstrom P, Hormuzdiar W, Milenkovic M, Aglieco F, Tyrrell L, Dib-Hajj S, Waxman SG, Black JA (2000) Localization of the tetrodotoxin-resistant sodium channel NaN in nociceptors. *Neuroreport* 11: 199–202

67 Liu C-J, Dib-Hajj SD, Black JA, Greenwood J, Lian Z, Waxman SG (2001) Direct interaction with contain targets voltage-gated sodium channel Na$_V$1.9/NaN to the cell membrane. *J Biol Chem* 276: 46553–46561

68 Black JA, Waxman SG (2002) Molecular identities of two tetrodotoxin-resistant sodium channels in corneal axons. *Exp Eye Res* 75: 193–199

69 Herzog RI, Cummins TR, Waxman SG (2001) Persistent TTX-resistant Na$^+$ current affects resting potential and response to depolarization in simulated spinal sensory neurons. *J Neurophysiol* 86(3): 1351–1364

70 Baker MD, Chandra SY, Ding Y, Waxman SG, Wood JN (2003) GTP-induced tetrodotoxin-resistant Na$^+$ current regulates excitability in mouse and rat small diameter sensory neurones. *J Physiol (Lond)* 548: 373–382

71 Carlton JM, Dougherty PM, Pover CM, Coggeshall RE (1991) Neuroma formation and numbers of axons in a rat model of experimental peripheral neuropathy. *Neurosci Lett* 131: 88–92

72 Basbaum AI, Gautron M, Jazat F, Mayes M, Guibaud G (1991) The spectrum of fiber loss in a model of neuropathic pain in the rat: an electron microscopic study. *Pain* 47: 359–367

73 Xie Y, Zhang, J, Petersen M, LaMotte RH (1995) Functional changes in dorsal root ganglion cells after chronic nerve constriction in the rat. *J Neurophysiol* 72: 466–470

74 Zhang JM, Song XJ, LaMotte RH (1997) An *in vitro* study of ectopic discharge generation and adrenergic sensitivity in the intact, nerve-injured rat dorsal root ganglion. *Pain* 72: 51–57

75 Dib-Hajj SD, Fjell J, Cummins TR, Zheng Z, Fried K, LaMotte R, Black JA, Waxman SG (1999) Plasticity of sodium channel expression in DRG neurons in the chronic constriction injury model of neuropathic pain. *Pain* 83: 591–600

76 Hains BC, Saab CY, Klein JP, Craner MJ, Waxman SG (2004) Altered sodium channel expression in second-order spinal sensory neurons contributes to pain after peripheral nerve injury. *J Neurosci* 24: 4832–4839

77 Kral MG, Xiong Z, Study RE (1999) Alteration of Na$^+$ currents in dorsal root ganglion neurons from rats with a painful neuropathy. *Pain* 81(1–2): 15–24

78   Basbaum AI, Wall PD (1976) Chronic changes in the response of cells in adult cat dorsal horn following partial deafferentation: the appearance of responding cells in a previously non-responsive area. *Brain Res* 116: 181–204

79   Woolf CJ (1983) Evidence for a central component of post-injury pain by persensitivity. *Nature* 306: 686–688

80   Woolf CJ, Wall PD (1986) Relative effectiveness of C primary afferent fibers of different origins in evoking a prolonged facilitation of the flexor reflex in the rat. *J Neurosci* 6: 1433–1442

81   Laird JM, Bennett GJ (1993) An electrophysiological study of dorsal horn neurons in the spinal cord of rats with an experimental peripheral neuropathy. *J Neurophysiol* 69: 2072–2085

82   Craner MJ, Klein JP, Renganathan M, Black JA, Waxman SG (2002) Changes of sodium channel expression in experimental painful diabetic neuropathy. *Ann Neurol* 52: 786–792

83   Hong S, Morrow TJ, Paulson PE, Isom LL, Wiley JW (2004) Early painful diabetic neuropathy is associated with differential changes in tetrodotoxin-sensitive and -resistant sodium channels in dorsal root ganglion neurons in the rat. *J Biol Chem* 279(28): 29341–29450

84   Wuarin-Bierman L, Zahnd GR, Kaufmann F, Burcklen L, Adler J (1987) Hyperalgesia in spontaneous and experimental animal models of diabetic neuropathy. *Diabetologia* 30(8): 653–658

85   Ahlgren SC, Levine JD (1993) Mechanical hyperalgesia in streptozotocin-diabetic rats. *Neuroscience* 52: 1049–1055

86   Courteix C, Eschalier A, Lavarenne J (1993) Streptozotocin-induced diabetic rats: behavioral evidence for a model of chronic pain. *Pain* 53: 81–88

87   Calcutt NA, Jorge MC, Yaksh TL, Chaplan SR (1996) Tactile allodynia and formalin hyperalgesia in streptozotocin-diabetic rats: effects of insulin, aldose reductase inhibition and lidocaine. *Pain* 68: 293–299

88   Tanaka M, Cummins TR, Ishikawa K, Black JA, Ibata Y, Waxman SG (1999) Molecular and functional remodeling of electrogenic membrane of hypothalamic neurons in response to changes in their input. *Proc Natl Acad Sci USA* 96: 1088–1093

89   Cummins TR, Black JA, Dib-Hajj SD, Waxman SG (2000) Glial-derived neurotrophic factor upregulates expression of functional SNS and NaN sodium channels and their currents in axotomized dorsal root ganglion neurons. *J Neurosci* 20: 8754–8761

90   Waxman SG, Cummins TR, Dib-Hajj S, Fjell J, Black JA (1999) Sodium channels, excitability of primary sensory neurons, and the molecular basis of pain. *Muscle Nerve* 22: 1177–1187

91   Black JA, Liu S, Tanaka M, Cummins TR, Waxman SG (2004) Changes in the expression of tetrodotoxin-sensitive sodium channels within dorsal root ganglia neurons in inflammatory pain. *Pain* 198: 237–247

Current approaches for the discovery of novel Na$_V$ channel inhibitors for the treatment of brain disorders

Jeffrey J. Clare

Gene Expression and Protein Biochemistry Department, GlaxoSmithKline, Stevenage, Hertfordshire SG1 2NY, UK

Introduction

Na$_V$ channel inhibitors are an important class of drugs that are used to treat a number of CNS indications including pain, local anaesthesia, epilepsy and bipolar disorder. Despite the indispensable role that Na$_V$ channels play within the CNS, these drugs are considered safe and have few side effects [1]. In sharp contrast to the lethal effects of simple open channel blockers (e.g., tetrodotoxin), Na$_V$ inhibitory drugs are tolerable due to their remarkably subtle modulation of channel function. This is generally thought to be because their effects are both use- and voltage-dependent – that is, the extent of inhibition depends on the rate of channel firing and on the membrane potential (Fig. 1). These properties are also highly important for therapeutic efficacy since the extent of channel block is greatly increased during periods of repetitive firing or sustained depolarisation as can occur, for example, during seizure activity or during pain signalling.

Na$_V$ inhibitory drugs in current use were discovered empirically using traditional pharmacological approaches, and were only subsequently found to inhibit Na$_V$ channels. Despite the cloning and characterisation of a multi-gene family of Na$_V$ channels [2] the development of improved Na$_V$ blockers (e.g., with greater potency and/or selectivity) using molecular target-driven approaches has proven extremely difficult. Compared to more tractable targets of pharmaceutical interest (e.g., ligand gated channels, G-protein coupled receptors, enzymes), these channels are relatively recalcitrant to the methods typically used in the drug discovery process. This is partly due to the considerable technical challenges involved in manipulating and expressing these large, unstable and sometimes toxic genes. Additionally, the nature of these proteins necessitates the use of relatively complex functional assays, which are challenging to configure in a format consistent with the high throughput required for large-scale random screening. Furthermore, given the mechanism of action outlined above, the ability to monitor subtle parameters, such as voltage and/or use-dependence, at a sufficiently large scale for secondary screening is crucial to success but is also extremely difficult.

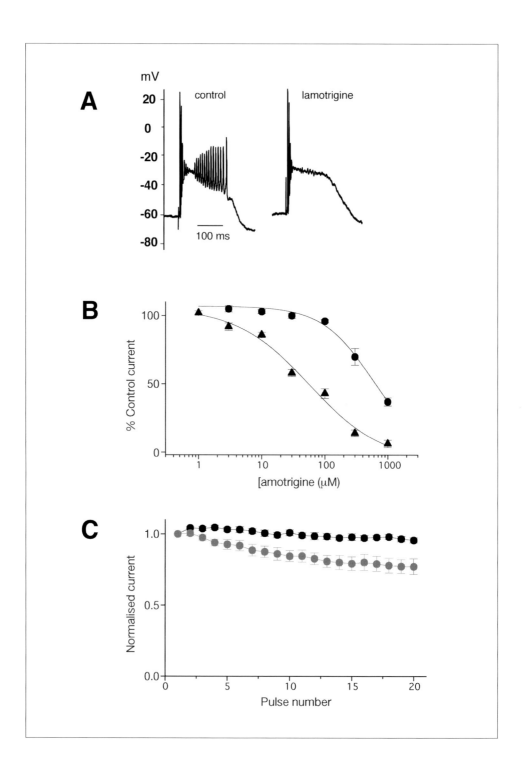

Despite these formidable challenges, significant technical advances have been made in recent years, which have dramatically increased the prospects of developing improved therapeutic Na$_V$ blockers. In parallel with this, an ever-increasing body of evidence linking Na$_V$ channels with CNS diseases is being uncovered. The aims of this chapter are to review the recent advances in these two aspects of Na$_V$ research which, together, are re-invigorating Na$_V$ drug discovery within the pharmaceutical industry.

Na$_V$ channels as therapeutic targets for CNS diseases

Na$_V$ blocking drugs arose as long ago as 1905 with the discovery of synthetic amino benzoates, such as procaine, which have local anaesthetic effects. These compounds were later discovered (1936) to prevent cardiac arrhythmias when locally applied during heart surgery. At about this time phenytoin, a non-sedating analogue of phenobarbitol, was first synthesised and found to suppress electrically-induced seizures in animals. However, it was not until much later that these drugs were shown to modulate Na$_V$s at clinically relevant concentrations – first procaine-related local anaesthetics (1959) and subsequently phenytoin (1983). These drugs were the forerunners of other Na$_V$ blockers, e.g., carbemazepine and lamotrigine, which have subsequently strengthened and extended the rationale for Na$_V$s as therapeutic targets for local anaesthesia, cardiac arrhythmia, epilepsy, and a variety of other diseases. The interpretation of these findings is somewhat complicated since these drugs can also exhibit effects on other targets (including, for example, other volt-

Figure 1
Voltage and use-dependent action of Na$_V$ inhibitors
A) Lamotrigine prevents repetitive firing of action potentials in an in vitro *model. Synaptically evoked action potentials were measured by intracellular recordings from neurons in rat hippocampal slices superfused with normal artificial cerebrospinal fluid containing bicuculline (20 µM) but without Mg^{2+}. Sustained repetitive firing is observed which is blocked by lamotrigine (50 µM). B) Potency of lamotrigine inhibition of hNa$_V$1.2 stably expressed in CHO cells is voltage dependent. Concentration-response curves were generated by measuring currents evoked by depolarising pulses to 0 mV from holding potentials (V$_h$) of either -90 mV (●) or -60 mV (■) in the presence of different concentrations of lamotrigine. IC$_{50}$ values were 641 µM and 56 µM at V$_h$ of –90 and –60 mV respectively. C) Inhibition of hNa$_V$1.2 by lamotrigine is use-dependent. Trains of depolarising pulses (20 msec duration, 10 Hz) from a V$_h$ of –90 mV were applied and the currents elicited by each pulse normalised to the first pulse to remove the effects of tonic block. Inhibition by lamotrigine (100 mM, grey circles) progressively increases with each additional pulse. (Reproduced from [72, 111] with permission from Springer-Verlag)*

age-gated ion channels [3, 4]). Thus, other mechanisms of action may contribute to efficacy or influence the distinctive profiles of these inhibitors with respect to different indications. The cloning of the first Na_V cDNA [5] and the subsequent expansion of the multi-gene Na_V family have set the scene for molecular and genetic approaches to help analyse this issue. This section is aimed at reviewing some of the more recent genetic, molecular and pharmacological evidence that suggests a role for Na_V channels in selected CNS diseases.

Na_V channels and epilepsy

Epilepsy is a group of brain disorders characterised by the common symptom of recurrent paroxysmal seizures. These arise from unprovoked, synchronised burst-firing of action potentials in neuronal populations within the brain. Thus, given the role of Na_Vs in generating and propagating action potentials, and their general role in neuronal excitability, it is not surprising that they have been found to be important both in the pathogenesis and in the treatment of epilepsy.

Reductionist approaches to studying seizures, e.g., using brain slices or simple neuronal circuits that mimic burst-firing activity, have confirmed the importance of Na_Vs in this process (Fig. 1a, [6]). In such models, pro-convulsant agents enhance sodium-dependent action potentials either by prolonging them, e.g., like pentylenetetrazole (PTZ) [7], or by increasing peak sodium conductance, e.g., like cocaine [8]. Conversely, agents that increase the open probability of Na_Vs, e.g., veratridine [9] or pyrethroids [10] are epileptogenic in these models. Furthermore, Na_V channel blocking anticonvulsant drugs such as phenytoin [11] or lamotrigine (Fig. 1a) prevent burst-firing activity even when this is induced by mechanisms not directly related to Na_Vs. These drugs are thought to have this effect by preferentially binding to, and stabilising, an inactivated state of the channel, which is consistent with the voltage-and use-dependent mechanism of action discussed above.

In addition to inhibiting the transient Na^+ currents that underlie action potentials, another potential anticonvulsant mechanism might be blockade of non-inactivating or persistent Na^+ currents (INaP). INaP has been found to occur in a variety of neuronal types although it normally makes up only a small proportion of total peak Na^+ current (1–3%) in these cells. Nevertheless, this could be functionally significant since this causes inward currents to occur at membrane potentials when transient channels are inactive. Thus, INaP probably plays a crucial role in neuronal excitability, e.g., by modulating resting membrane potential, and its importance is increasingly being recognised in regulating the integration of synaptic input and output, shaping repetitive firing and the generation of neuronal rhythmicity. The precise molecular identity of Na^+ channel subtypes that cause persistent currents, and the mechanism involved, are not yet fully elucidated though INaP can be observed in recombinant Na_V channels (Fig. 5, more fully discussed below). INaP occurs at high levels in subicular neurons from human temporal lobe epilepsy (TLE) patients, and can be inhibited by Na$_V$-blocking anticonvulsants like phenytoin [12, 13], topiramate [14] and lamotrigine [15]. A causative role in epilepsy for INaP has now been implicated by studies of Na$_V$-linked human epilepsy mutations [16]. This is also supported by studies showing transgenic mice with increased INaP due to the expression of an inactivation impaired mutant Na$_V$1.2 exhibit a severe epilepsy phenotype [17].

As indicated above, the study of mutations associated with inherited forms of the epilepsy has further confirmed the role of Na$_V$s in this disease and has also given insight into the kind of alterations in channel function that can lead to seizure activity. While this field was slow to take off, since the discovery and characterisation of the first Na$_V$-linked human epilepsy mutation in 1998 [18], a number of mutations have now been identified and studied. To date, these have been discovered in three different Na$_V$ genes, SCN1A, SCN2A and SCN1B, and are linked to three different epilepsy syndromes: generalised epilepsy with febrile seizures plus (GEFS+ types 1 and 2), severe myoclonic epilepsy of infancy (SMEI), and benign familial neonatal-infantile seizures (BFNIS).

Mutations in SCN1A can cause GEFS+, a relatively benign syndrome, or SMEI, which is much more severe. Interestingly, there appears to be a correlation between the nature of the mutation and the severity of the symptoms of the corresponding disease. That is, truncating and missense mutations in the pore-forming regions (S5–S6) nearly always lead to a classical form of SMEI, whereas missense mutations in the voltage sensor (S4) can lead to milder SMEI or GEFS+, and missense mutations outside S4–S6 region mostly lead to GEFS+, or occasionally to milder forms of SMEI [19]. The effects of a number of different SCN1A missense mutations on Na$_V$ channel function has been examined in various heterologous systems. Probably the most definitive studies are those in which the mutations have been introduced into the cloned human Na$_V$1.1 and then expressed in mammalian cells [16, 20, 21]. Interestingly, both nonsense (R712X, R1407X, R1892X) and missense (G979R, L986F, F1831S) SMEI mutations were found to abolish channel function, consistent with the idea that loss of function and haploinsufficiency for SCN1A causes SMEI. In addition, in this system, three different GEFS+ mutations (R1648H, T875M, W1204R) were found to cause "gain-of-function" effects, altering channel inactivation and leading to persistent inward sodium currents (INaP) suggesting a highly plausible disease mechanism. However, other GEFS+ mutations tested did not cause this "gain-of-function" effect; two of them resulted in more subtle defects (I1656M, R1657C) and two others abolished function altogether (V1353L, Λ1685V).

GEFS+ has also been found to be linked to a mutation in SCN2A [22] although recent evidence suggests SCN2A mutations more commonly cause BFNIS [23]. Functional analysis of the SCN2A GEFS+ mutation (R187W) when introduced into rat Na$_V$1.2 indicates it causes slowed inactivation of the channel potentially leading to increased Na^+ influx and excitability *in vivo* [22]. Only one nonsense mutation (R102X) has been reported for SCN2A but, consistent with the trend observed for SCN1A, this is associated with a more severe form of intractable epilepsy that is related to, but distinct from, SMEI [24]. There is evidence to suggest that the truncated fragment resulting from this mutation may have a dominant negative effect which would further exacerbate haploinsuffiency of the $Na_V1.2$ protein.

Mutations in SCNB1 also cause GEFS+. Two different alterations have been described, a missense mutation (C121W) [18] and a 5 amino acid deletion (I70ΔE74) [25] that both occur in the Ig-like fold of the extracellular domain. The C121W mutation has been analysed *in vitro* by expression in a mammalian cell background and is found to cause loss of function of the β subunit. This has only relatively subtle effects on channel function, but leads to a greater proportion of channels that are available to open at hyperpolarised potentials and during high frequency activity [26]. In summary, investigation of inherited Na_V-linked epilepsies has given insight into the kind of alterations in channel function that can lead to seizure activity. Evidence to date tends to suggest the severity of the observed epilepsy phenotype may be related to the severity of effect the mutation has on channel function. However, there are obvious exceptions to this trend indicating that, as with most channelopathies, extrapolating *in vivo* and cellular consequences from the biophysical defects observed *in vitro* is not necessarily straightforward.

Since there is relatively strong rationale for involvement of Na_V channels in epilepsy, it has been postulated that alterations in Na_V expression or functional properties may underlie idiopathic forms of the disease. Thus, numerous studies have attempted to document differences in expression and function in disease *versus* normal tissue. An obvious limitation with such correlations is that it is not clear whether any changes observed are causative or are consequential to the epileptic condition or to drug treatment. Given the inherent difficulties in addressing this in humans (discussed further below), a number of investigators have approached this question by studying animal models. Analysis of different Na_V subtype α subunit mRNAs following kainic acid-induced seizures revealed a marked increase in $Na_V1.3$ expression and a modest increase in $Na_V1.2$ expression in rat hippocampus [27]. In the same model, it was also shown that seizure activity is correlated with alternative splicing of $Na_V1.2$ and $Na_V1.3$ mRNAs, causing an increase in the presence of the neonatal isoforms in the adult hippocampus [28]. A similar phenomenon was observed in a post-status epilepticus (electrically induced) model of chronic epilepsy [29]. Increased expression of neonatal $Na_V1.2$ and $Na_V1.3$ splice variants were observed in CA1–CA3 and dentate gyrus (DG) neurons, which persisted for at least three months after status epilepticus (SE), corresponding to the chronic seizure phase. Functional analysis of Na_V currents showed a negative shift in voltage dependence of activation leading to an increased window current in CA1 (though not in DG) neurons from similar post-SE animals

[30]. It is possible such changes are due to the increased expression of the neonatal splice variants, though this remains to be directly proven. A similar phenomenon has been reported in a pilocarpine-induced post-SE model. In this case the increased window current was observed in DG neurons and was due to a positive shift in inactivation as well as a negative shift in activation [31]. In this study neonatal splice variants were not measured, though persistent decreases were observed in Na$_V$1.2, Na$_V$1.6, Na$_V$β1 mRNAs and a transient decrease in Na$_V$β2 mRNA was found. Interestingly, in this model, there is evidence to suggest the pharmacology of Na$_V$ currents in DG cells is also altered [32, 33]. Use-dependent block by carbemazepine is abolished compared to control; and tonic block by carbemazepine, lamotrigine and phenytoin is reduced while the slowing of recovery from fast inactivation by carbemazepine and phenytoin are reduced. In a related model (lithium/pilocarpine induced post-SE), an increase in INaP was observed in entorhinal cortex layer V neurons [34]. This was due to selective upregulation of non-inactivating current rather than enhanced window current resulting from the increased overlap of activation and inactivation curves.

Given the limited availability of suitable human brain tissue samples, it is not surprising that fewer studies have been carried out to investigate corresponding changes in human epilepsy. While it is possible to obtain tissue that has been surgically resected from the temporal lobe of adult patients with intractable epilepsy, there are inherent caveats associated with interpreting the data from studies using such material. The biggest problem is the obvious lack of directly comparable control tissue. Adjacent tissues that are also removed with the hippocampus, e.g., temporal cortex, do not necessarily provide reliable controls since changes in gene expression may also occur within these regions [35]. Post-mortem tissue can be used for comparison in gene expression (but not functional) studies, though it is not always possible to adequately control for factors such as patient age, differences in tissue processing and preservation, stability of mRNA in post-mortem *versus* post-surgical tissue etc. In addition, since resections are normally only carried out following extensive pharmacotherapy, by default, any changes that are observed could be related to the drug treatment(s) rather than to the disease itself.

Thus, very few studies documenting changes in Na$_V$ expression associated with human epilepsy have been reported. In one study, using a novel but not well established assay, an increase in the ratio of Na$_V$1.1 and Na$_V$1.2 mRNA was measured in hippocampi resected from temporal lobe epilepsy patients compared to post-mortem control tissue, though it was not clear if this was due to a decrease in Na$_V$1.1 or an increased in Na$_V$1.2 or both [36]. In a more detailed study, using *in situ* hybridisation, a decrease in Na$_V$1.2 mRNA in the CA1-3 sub-fields of the hippocampus and an increase in Na$_V$1.3 mRNA in the CA4 subfield were observed (Fig. 2) [37]. No changes were observed in Na$_V$1.1 or Na$_V$1.6.

Despite the logistical and technical challenges in studying functionally active human neuronal preparations, there a number of reports characterising Na$_V$ cur-

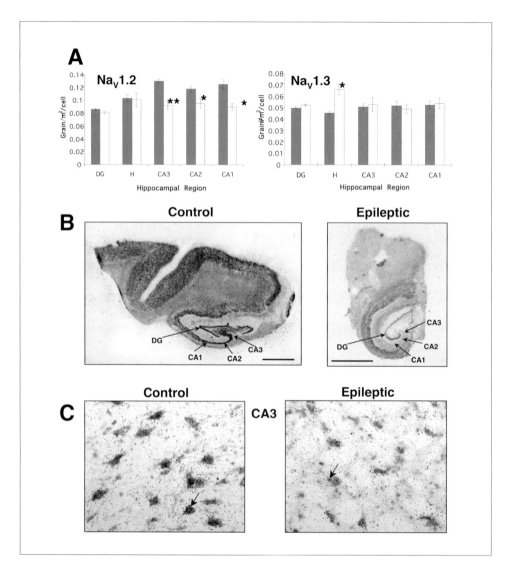

Figure 2

Changes in Na$_V$ channel mRNA levels in human epilepsy

A) Significant down-regulation of Na$_V$1.2 mRNA in regions CA1-3 are observed in hippocampus, whereas up-regulation of Na$_V$1.3 mRNA is observed in CA4 (hilus). B) Macroscopic image of hippocampus from post-mortem control (left) and epileptic (right) human brain showing reduced expression of Na$_V$1.2 mRNA in CA1-3. C) Higher magnification image showing cellular distribution of Na$_V$1.2 mRNA and reduced staining in the CA3 region of epileptic (right) human hippocampus compared to post mortem control (left) (Reproduced from [37] with permission from Elsevier)

rents from human epileptic tissue [38, 39]. Their functional properties are similar to those described for animal preparations, though current densities were found to be comparatively high in epileptic hippocampal neurons [39] and a very high proportion INaP (up to 53% of total Na$^+$ current) was observed in subicular neurons from epilepsy patients [40]. Importantly, when compared to DG neurons from control rats, Na$^+$ currents in human epileptic DG neurons showed similar differences in sensitivity to carbamazepine as those from pilocarpine-treated rats, i.e., use-dependent block did not occur [41]. This insensitivity to carbamazepine occurred only in neurons from patients that were clinically insensitive to carbamazepine prior to surgery, suggesting the intriguing possibility that this may be a possible mechanism of pharmacoresistance [33].

Na$_V$ channels and bipolar disorder

Bipolar disorder (BPD) is a severe chronic illness characterised by two apparently opposite mood states, mania and depression. The manic phase is characterised by hyperarousal, (increased motor activity, racing thoughts, impaired judgement, decreased sleep) whereas the depressive phase has similar symptoms to major depression (depressed mood, cognitive and psychomotoric changes etc.). The aetiology and pathophysiology of the disease are only beginning to be understood (for reviews see [42, 43]). Progress in this area, and also in the development of new therapeutic agents, has been hampered by the lack of suitable animal models that mimic all the key features of BPD, i.e., both the manic and depressive phases as well as cyclicity of the disease [44]. Early hypotheses focussed on the role of monoamines, since depression has been conceptualised as a deficiency in certain monoaminergic systems and many antidepressants that increase the activity of these can precipitate mania. Other potential mechanisms that have been more recently suggested include disturbances in other neurotransmitter systems (including glutamatergic), alterations in intracellular signal transduction pathways, impairment of neuronal plasticity, and maladaptation of the stress related hypothalamus–pituitary–adrenal pathway (see [42]). Given the lack of animal models, these hypotheses are derived mainly from functional and morphological findings from non-invasive neuroimaging studies or inferred from analysing the potential mechanisms of action of agents that are used to treat the disease.

Currently there are three main types of medication used for therapy of BPD; lithium, anticonvulsants and antipsychotics. The effectiveness of Na$_V$-blocking anticonvulsants such as carbamazepine and lamotrigine has implicated the involvement of these channels in the disease [45, 46]. However, since these agents also have activity at other targets, more conclusive proof awaits the development of Na$_V$ inhibitors with increased selectivity. If efficacy in BPD is indeed directly related to Na$_V$` inhibition, it is not clear whether the mechanism involved might be similar to that proposed for anti-seizure activity, i.e., *via* selective blockade of repetitive high frequency firing. One possible explanation that is consistent with this is provided by the kindling model of BPD [47]. This model draws an analogy with the process of limbic kindling in rodents, whereby electrical stimulation of limbic structures initially elicits no response but upon repetition causes seizures that eventually become spontaneous. Na_V blockers such as carbemazepine and lamotrigine are effective in preventing kindled seizures (see [48]) and some may also prevent the development of the kindling process [49]. In BPD it is postulated that stress, or other psychosocial triggers, could lead to an analogous kindling or sensitisation in susceptible individuals causing manic and depressive episodes and a progressive worsening of symptoms that eventually results in these episodes occurring even in the absence of such a trigger. It has also been suggested that this process involves structural and functional changes. In support of this there is evidence, from positron emission tomography (PET), for alterations in regional cerebral blood flow and glucose metabolism in limbic and prefrontal cortex structures in patients with mood disorders, and also for corresponding morphological changes as measured by magnetic resonance imaging (MRI, see [50] for review). By analogy with limbic kindling in animals, these structural and functional changes in BPD could lead to limbic hyperexcitability, which would be blocked by Na_V inhibitors leading to amelioration of the condition [4].

Additional mechanisms of Na_V blockers may also be relevant to BPD, for example, modulation of glutamate release and neuroprotection (see [48]). These effects are highly relevant for the treatment of cerebral ischaemia and so will be discussed in more detail in the next section. However, one interesting line of investigation that is cogent for BPD is the phenomenon known as pre-pulse inhibition (PPI), first described as long ago as 1939 [51] but more systematically studied in the 1960s [52]. PPI is where the reflex startle reaction elicited by a sudden noise is reduced if preceded by a low-level pulse of background noise. This effect is remarkably robust, occurs in humans as well as animals, and has been used as a cross-species model of sensorimotor gating (see [53]). Furthermore, PPI is impaired in certain neuropsychiatric conditions including schizophrenia [54], obsessive-compulsive disorder [55], Tourette's syndrome [55] and acute mania [56], though this deficit is not apparent in patients who respond to antipsychotic drugs. Glutamatergic systems are involved since NMDA receptor antagonists like ketamine cause deficits in PPI, both in primates and in rodents [57, 58]. Interestingly, lamotrigine prevents ketamine-induced PPI in mice and also enhances PPI in the absence of these agents [59]. Most likely these effects are exerted *via* the glutamatergic system and inhibition of glutamate release, since lamotrigine does not prevent amphetamine-induced deficits. In agreement with this, lamotrigine also decreases the perceptual abnormalities caused by ketamine in humans [60]. Further studies are required, but it is possible these effects of lamotrigine on PPI are pertinent to its efficacy in BPD, particularly in delaying the onset of the manic phase.

Na$_V$ channels and cerebral ischaemia

Cerebral ischaemia is the interruption of blood flow within the brain, which leads to both acute and delayed neuronal damage, often manifested as stroke. The energy deprivation caused by brain ischaemia triggers synaptic and cellular events that are somewhat similar to that occurring during seizures, where the maintenance of a sustained depolarised state in the hyperactive neuron also compromises its energy supply. Thus, a number of anti-epileptic drugs have been tested as possible neuroprotective agents in animal models of stroke and experimentally induced ischaemia (reviewed in [61]). These studies suggest therapeutic strategies that reduce excessive release of excitatory neurotransmitters or enhance neuronal inhibition might be beneficial. However, while many of these agents show good efficacy in such models, results have so far been generally disappointing in clinical studies. Nevertheless, the potential neuroprotective role of anti-epileptic agents remains an area of high interest in the search for anti-ischaemic drugs.

Various anticonvulsant Na$_V$ channel inhibitors have been found to be neuroprotective in experimental models of ischaemia (reviewed in [62]). As was noted previously in the discussion of BPD, the interpretation of these results is complicated by the fact that these inhibitors also have activities at other targets. Nevertheless, there is good rationale for beneficial effects of Na$_V$ channel blockade. A key part of this argument is that reduction of Na$^+$ influx is likely to have an ATP sparing effect. This is because, in neurons, most ATP is used to fuel the Na$^+$/K$^+$ ATPase pump that normally maintains the electrochemical gradient for Na$^+$. A combination of excitotoxic stimulation and high rates of action potential firing would cause excessive Na$^+$ loading in neurons around the ischaemic region, with consequent ATP depletion due to overactivity of the Na$^+$/K$^+$ pump. ATP depletion is known to be an early event during brain ischaemia [63]. In addition to conserving ATP itself, prevention of Na$^+$ overloading could have other protective effects. A further consequence is that, as ATP is depleted, the Na$^+$/K$^+$ pump becomes inactive causing run down of the electrochemical gradient. This, in turn, causes inactivity of the Na$^+$/Ca^{2+} antiporter that normally extrudes Ca^{2+} from neurons thus contributing to the accumulation of cytoplasmic Ca^{2+}. This process also occurs in presynaptic terminals and in glial cells and, in combination with membrane depolarisation, leads to release of excitotoxic levels of glutamate. Furthermore, Na$^+$ dependent glutamate re-uptake transporters at the synapse and in adjacent glial cells may actually be reversed, due to the high intracellular levels of Na$^+$, causing massive levels of glutamate to accumulate. This causes activation of NMDA and metabotropic glutamate receptors in neurons causing Ca^{2+} overload and triggering a cascade of events leading to tissue damage, cell swelling, activation of proteolytic enzymes, generation of free radicals and eventually cell death. Thus, prevention of neuronal depolarisation and Na$^+$ overloading using Na$_V$ inhibitors is potentially an important step at which to break this vicious cycle.

Despite support for this rationale and for efficacy of Na_V inhibitors in animal models (see [62]), as with other types of anticonvulsant, this has yet to be success-fully translated into a clinical setting. Sipatrigine (a lamotrigine analogue with Na_V channel blocking as well as additional activities see [3]), crobenetine and fospheny-toin (a pro-drug of phenytoin) have all been evaluated in Phase II and/or Phase III clinical trials, but their development has been discontinued. Several possible expla-nations may underlie this paradox, including clinical trial design, heterogeneity within the stroke population, lack of sensitivity of outcome measurements in human studies, limited temporal therapeutic window, or pathophysiological differences between animals and humans. Further investigation of the mechanisms underlying human stroke *versus* experimental models, plus a better understanding of any potential differences between them, is needed to aid progress in this area. To this end the development of more selective Na_V channel inhibitors would also be of benefit. It is also possible that additional therapeutic strategies, used in combination with anticonvulsant Na_V inhibitors, are required to more effectively target the "delayed" apoptotic neuronal death caused by human brain ischaemia.

Current approaches for the discovery of novel Na_V channel inhibitors

As described above, currently used therapeutic Na_V inhibitors all have their origins in traditional "empirical" pharmacology and it was only some time after their dis-covery that they were found to inhibit Na_V channels. The discovery, cloning and characterisation of a multigene family of Na_V channels has paved the way for more rational and specifically targeted approaches to finding therapeutic Na_V channel inhibitors. It is anticipated these approaches will lead to improved inhibitors with increased potency and selectivity that should enable further insight to be gained into the physiological and pathophysiological roles of Na_Vs and ultimately to the devel-opment of more effective therapies.

Cloning and analysis of human brain Na_V channels

The first Na_V channel cDNA to be isolated was cloned in 1984 by taking advantage of the abundant expression of these channels in the electric organ of the electroplax eel, *Electrophorus electricus* [5]. This ultimately led to the cloning and identification of an unexpectedly large gene family that now consists of nine highly related but dis-tinct subtypes, of which four are normally expressed in brain, $Na_V1.1$–1.3 and $Na_V1.6$ (Fig. 3). In addition, an "atypical" channel (Na_X) has been identified which, although related by sequence, may be functionally distinct and probably represents a different subfamily [2, 64, 65]. The cloning of the first mammalian Na_V cDNAs ($Na_V1.1$–1.3, [66, 67]) precipitated numerous studies aimed at molecular charac-

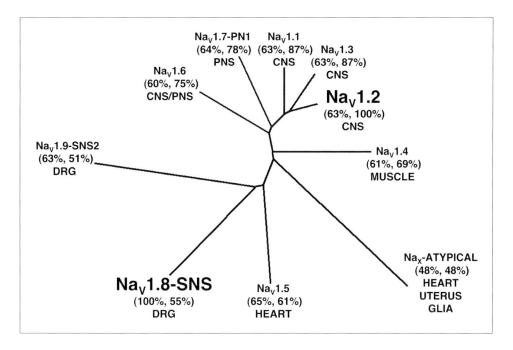

Figure 3
Phylogenetic tree for human voltage-dependent sodium channels
The primary tissue(s) in which they are expressed and the percentage amino acid sequence identities relative to representative tetrodotoxin- sensitive (Na$_V$1.2, 1st value) and insensitive (Na$_V$1.8, 2nd value) subtypes are indicated.

terisation of Na$_V$ channels and knowledge of their structure–function relationships is now extensive (for review see [68]). However, information relating to brain channels has largely come from the study of the rat Na$_V$1.2 isoform which has been considered a prototypical brain subtype.

Until recently, relatively few studies have characterised the other brain subtypes from rat and even less information has been available for the human orthologues. This partly reflects the technical challenges in handling Na$_V$ channels and their cDNAs which makes them difficult to manipulate and heterologously express. Although a comprehensive and direct comparison of the functional properties of the Na$_V$ channel subtypes has not yet been published, on collation of data from various labs only relatively subtle differences between the various rodent Na$_V$ orthologues are apparent (reviewed in [69]). In one of the few direct comparisons reported, small differences in activation (voltage dependence) and inactivation (voltage dependence and kinetics) were noted for Na$_V$1.6 compared to Na$_V$1.1 and Na$_V$1.2 but these differences were not significant if β1 and β2 subunits were co-expressed [70]. Potentially important differences were observed in the level of INaP: $Na_V1.2$ gave the lowest proportion (~1% of total current under the conditions used), whereas at depolarised potentials $Na_V1.6$ mediated the most (~5%) and at hyperpolarised potential $Na_V1.1$ gave the most (~5%). That recombinant $Na_V1.6$ channels can mediate INaP is consistent with findings from $Na_V1.6$ null mice which show that the characteristic INaP normally observed in cerebellar Purkinje cells are absent in cells from these animals [71]. Nevertheless, these findings must be interpreted with caution since this was a cross-species comparison (murine $Na_V1.6$ *versus* rat $Na_V1.1$ and 1.2 orthologues) carried out using the *Xenopus* oocyte expression system, which may not be very representative of the situation mammalian neuronal cells.

Due to the importance of the human channels for drug discovery, and given the paucity of directly comparable data for the different subtypes, attention at Glaxo-SmithKline has been focussed on characterising the human Na_V orthologues. Thus, cDNAs for most of the human Na_V subtypes have been cloned and these have been stably expressed in mammalian cells [72–75, 112]. This has enabled analysis of the basic biophysical properties of the α subunit subtypes from human brain, allowing a comparison with published data for their rodent counterparts. Consistent with the high conservation of amino acid sequences, this comparison suggests the human and rodent orthologues are broadly similar in their basic properties, with only minor differences that may, at least in part, be due to the different recording conditions or expression systems used. Electrophysiological analysis of the human brain orthologues using exactly the same recording conditions and expressed in the same mammalian cell background (HEK293) has also allowed direct comparison of the four major brain subtypes for the first time. As found with the rodent orthologues, only relatively subtle differences between the subtypes can be observed (Fig. 4). Interestingly, the most distinctive subtype in this system is $Na_V1.2$, which inactivates at more depolarised potentials (V1/2 inact is 6–12 mV more positive) and recovers more rapidly from inactivation (τ_{inact} at the voltage giving maximum current is 2.6–3.4 fold less). While they appear to be rather subtle, these differences could have important consequences *in vivo* and, if also apparent with the native channels, would be expected to lead to greater availability of $Na_V1.2$ channels than the other subtypes during neuronal depolarisations. Along with distribution studies that indicate $Na_V1.2$ channels have a unique axonal localisation within the human brain (see below and Fig. 7) this may reflect a more specialised functional role for $Na_V1.2$ in the propagation of action potentials in unmyelinated neurons in contrast with the other brain subtypes.

A more striking difference between $Na_V1.2$ and the other brain subtypes is in the level of INaP observed when expressed in a human cell background (Fig. 5). Although the level of INaP observed does vary from cell to cell and from clonal cell line to clonal cell line, in HEK293 cells the $Na_V1.1$, 1.3 and 1.6 subtypes consistently mediate greater levels of INaP than $Na_V1.2$ (typically 1–40% of total peak

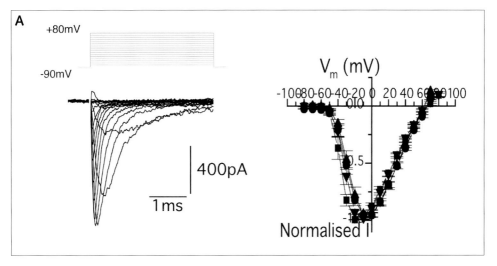

A

+80mV

-90mV

400pA

1ms

V_m (mV)

-100 80 -60 -40 -20 0 20 40 60 80 100

Normalised I'

B

	Na$_V$1.1	Na$_V$1.2	Na$_V$1.3	Na$_V$1.6
Activation V$_{1/2}$	−33 ± 1 mV	−24 ± 2 mV	−23 ± 3 mV	−29 ± 2 mV
Steady-state inactivation V$_{1/2}$	−72 ± 1 mV	−63 ± 1 mV	−69 ± 1 mV	−72 ± 2 mV
Recovery from inactivation	17 ± 3 ms	5 ± 1 ms	13 ± 2 ms	13 ± 2 ms
τ of inactivation (peak)	0.7 ± 0.1 ms	0.8 ± 0.0 ms	0.8 ± 0.1 ms	1.1 ± 0.1 ms

Figure 4
Comparison of the biophysical properties of human brain Na$_V$ subtypes
A) Left: representative traces showing depolarising pulses applied to HEK293 cells express-
ing human Na$_V$1.6 channels. Right: current-voltage relationships for the human brain NaV
subtypes – Na$_V$1.1 (■) Na$_V$1.2 (●), Na$_V$1.3 (▲) and Na$_V$1.6 (▼). B) Summary of the biophys-
ical parameters measured – data are presented as mean ± SEM (n = 4–7).

current for Na$_V$1.1, 1.3, 1.6, *versus* 0–5% for 1.2). The same trend is also observed when these subtypes are transiently expressed in HEK293 cells. The INaP observed in this system may be modulated *via* trimeric G-proteins, since the level of INaP decays when CsF is present in the recording pipette solution [74] and this is known to indirectly modulate G-protein activation [76]. G-protein involvement is consistent with other studies that show co-expression of Gβγ subunits in HEK293 cells induces increased levels of INaP mediated by rat Na$_V$1.2 [77] and, more recently, by human Na$_V$1.1 [112]. The finding that the different brain Na$_V$ subtypes have differing intrinsic basal levels of INaP could have profound implications for their roles

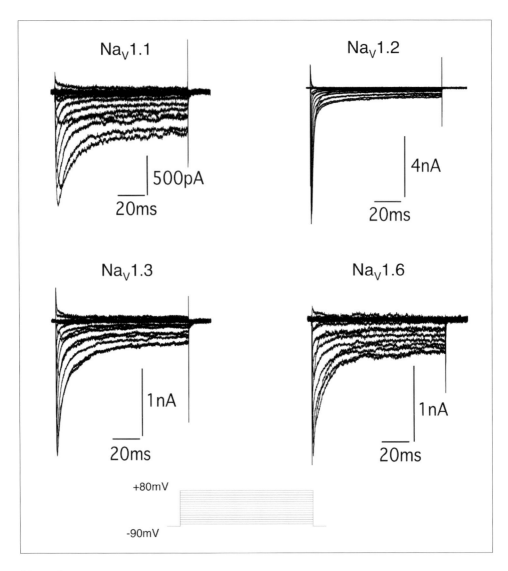

Figure 5
Currents mediated by human Na$_V$1.1, 1.3 and 1.6 subtypes show a prominent persistent component when expressed in a human cell background
Representative inward currents evoked by a series of depolarising pulses (100 msec duration, see bottom) are shown for all four brain subtypes when stably expressed in HEK293 cells. Currents decay with biphasic kinetics consisting of a rapid component (transient) and a sustained component (persistent). The sustained component persists for at least 100 msec and, with the exception of Na$_V$1.2, can comprise a large proportion the total current (up to 40% in some cells, depending on conditions used).

in vivo, both in normal and in pathological settings. A higher level of intrinsic INaP mediated by Na$_V$1.1, 1.3 and 1.6, together with their somato-dendritic localisation within brain neurons (Fig. 7), is consistent with these subtypes having a major influence on resting membrane potential in the cell body, as well as on processing of synaptic inputs, controlling frequency of firing, and on shaping burst firing behaviour, for example, during epileptiform hyperexcitability.

Cloning and expression of the human Na$_V$ subtypes has enabled the selectivity of existing therapeutic inhibitors to be profiled. It is conceivable that differences in relative efficacy of the various Na$_V$ inhibitors in different diseases and disease models may be related to differing potency at the different Na$_V$ subtypes. For example, a wide spectrum of *in vivo* efficacy in models of pain and seizure is observed in a series of structural analogues derived from lamotrigine [78]. However, no evidence for any appreciable intrinsic selectivity within the Na$_V$ family has been reported for any of the commonly used Na$_V$ inhibitors. For example, lamotrigine has similar potency for voltage dependent tonic block at each of the major brain subtypes (Fig. 6). Similarly, commonly used Na$_V$ inhibitors like lamotrigine appear to block the INaP mediated by recombinant channels with similar potency to the transient currents. A hint of improved subtype selectivity is beginning to emerge with more recently developed compounds [79] and, in addition, peptide toxins with subtype selective actions are beginning to be more fully characterised [80, 81]. The latter are potentially useful for exploring the specific physiological roles of the different subtypes as well as examining their relative importance in disease models.

Cloning of all the human Na$_V$ subtypes and the generation of comprehensive sequence information for the entire human Na$_V$ family has allowed the design and generation of a set of highly subtype-specific oligonucleotide probes and anti-peptide antibodies which can be used to map their distribution both in normal tissue as well in disease states [82–85]. These tools have been used to determine the pattern of Na$_V$ mRNA and protein expression in human brain. Until recently, most studies of this type have focussed on rodents [reviewed in 86] and relatively little information has been available for humans. Clearly, this is partly due to relative difficulty in obtaining suitable human tissue, although this also reflects the lack, until recently, of comprehensive sequence information for the human orthologues. In concordance with the highly conserved amino acid sequence and functional homology observed between rodent and human Na$_V$ orthologues, the overall distribution patterns of the Na$_V$ subtypes in human brain was found to be similar to that reported for rodents in the regions studied (cerebellum, somato-motor cortex, hippocampus, basal ganglia and thalamus [82, 83]). However, one striking difference is that Na$_V$1.3, which was previously considered from rodent studies to be mainly an embryonic or neonatally expressed subtype, is found to be widely expressed in the human adult brain (Fig. 7). In fact, subsequent immunological studies have now confirmed that this is also the case in rats [87].

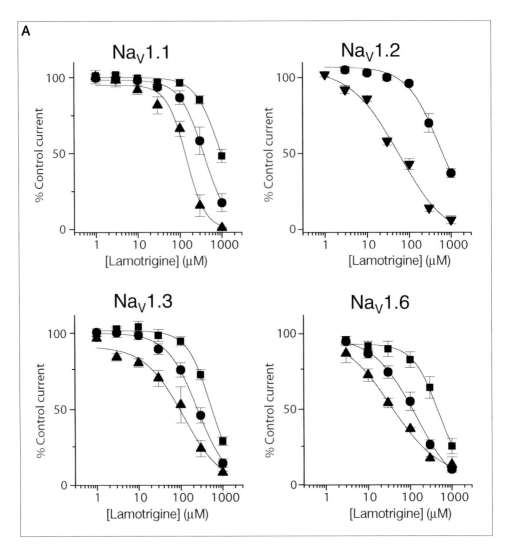

A

B

		Lamotrigine		TTX
	V_H −120 mV	V_H −90 mV	V_H −70 mV/−60 mV	
Na$_V$1.1	953 mM	377 mM	128 mM	6 nM
	[778–1167]	[231–614]	[81–205]	[5–7]
Na$_V$1.2	–	641 mM	56 mM	13 nM
Na$_V$1.3	545 mM	274 mM	116 mM	4 nM
	[478–622]	[203–368]	[65–208]	[4–5]
Na$_V$1.6	489 mM	127 mM	37 mM	3 nM
	[319–750]	[68–235]	[29–48]	[2–3]

In addition to confirming the mRNA distribution patterns obtained by *in situ* hybridisation, the human immunological studies have also confirmed and extended rodent studies of the sub-cellular localisation of the Na$_V$ subtypes in brain. The human Na$_V$1.2 subtype was shown to be uniquely concentrated along axons whereas the Na$_V$1.1, 1.3 and 1.6 were all found predominantly in neuronal cell bodies and proximal processes (Fig. 7). This pattern has important implications for their respective functions in the brain and strongly suggests a specialised role for Na$_V$1.2 in action potential propagation and a role in modulating synaptic inputs and outputs for the other subtypes. This may represent an important species difference since in rodents, in addition to Na$_V$1.2, Na$_V$1.6 protein was also found to be present in unmyelinated axons within the brain [88].

Recent advances in high-throughput assays for identifying Na$_V$ inhibitors

Over the last five years or so several technological advances have been made that are now revolutionising drug discovery for voltage-gated ion channels. These have dramatically improved the tractability of Na$_V$ channels to high-throughput random screening approaches. Clearly, patch-clamp electrophysiology has been the mainstay of ion channel research for decades and this is the "gold standard" assay since it allows full control over experimental conditions and provides exquisite sensitivity and temporal resolution. However, while being highly suitable as an analytical tool, the specialised, challenging and laborious nature of this technique, together with a severely limited throughput (e.g., 0–10 compound data points/day), are crucial disadvantages for its use in drug discovery. Thus, a "Holy Grail" within the pharmaceutical industry has been the development of automatable approaches that will allow the ability to make multiple parallel electrophysiological measurements, while at the same time "de-skilling" the process. In recent years a number of automated patch-clamp instruments potentially capable of achieving these goals have emerged (reviewed in [89]). Several of these systems are now commercially available and are already having a major impact on ion channel drug discovery.

One such system that has been successfully used for Na$_V$ currents is the Ion-Works instrument ([90], Molecular Devices Corporation). Similarly to several of the

Figure 6
Potency of inhibition by lamotrigine is similar for each human brain subtype
A) Concentration response curves for lamotrigine are shown for all four subtypes using three different holding potentials (–120 mV, –90 mV and –70 mV. N.B. Na$_V$1.2 was measured at –120 mV and –60 mV). B) Summary of IC$_{50}$ values expressed as geometric mean with 95% confidence limits (n = 4–7).

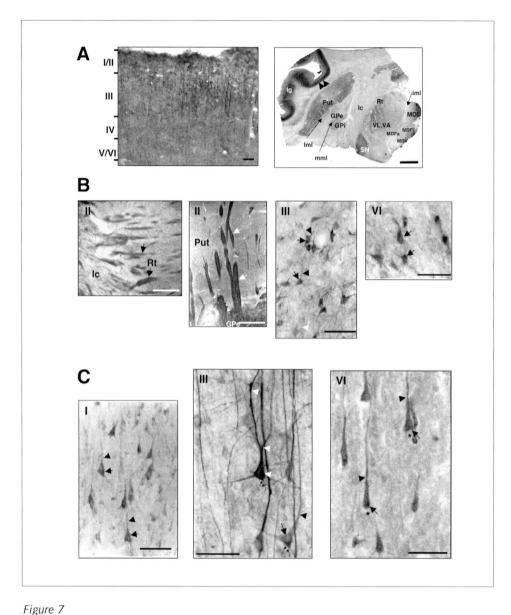

Figure 7

Immunolocalisation of Na$_V$ subtypes in adult human brain tissue

A) Na$_V$1.3 immunolocalisation in somato-motor cortex (left panel, roman numerals refer to the different cortical layers, scale bar = 200 µM) and basal ganglia/thalamus (right panel, Ig, insular gyrus; Put, putamen; GPe, external globus pallidus; GPi, internal globus pallidus; lml; external medullary lamina of the globus pallidus, mml, medial medullary lamina of the globus pallidus; SN, substantia nigra; Ic, internal capsule; Rt, reticular thalamic nucleus;

other automated systems, instead of a glass recording pipette this uses multiple recording apertures arranged in an array on a planar substrate. In the case of Ion-Works this takes the form of a plastic "patch plate" containing 384 wells, into each of which a centrally located microhole has been laser drilled. Unlike traditional patch-clamping, the cells are brought to the substrate, rather than bringing the pipette to the cells, by introducing a cell suspension into each well and then applying suction from below the hole. Once a seal between the cell and the substrate has formed, rather than sucking away the membrane bounded by the aperture as is done with other systems, electrical access is instead gained by permeabilising this patch of membrane with amphotericin B. This corresponds to the conventional perforated patch configuration and has the advantage that cellular co-factors that may modulate the channel under normal physiological conditions are more likely to be retained. The IonWorks instrument also benefits from a greater throughput than the other instruments currently available as it employs a 48 channel extracellular recording head and amplifier (the "intracellular" compartment of the cell in each well being connected to a common ground electrode). This enables 48 simultaneous recordings to be made (Fig. 8) and the entire 384 well plate to be read within minutes, following eight successive movements of the recording head. Depending on the voltage protocol being used, several thousand data points can be routinely generated per day representing an improvement in throughput of two orders of magnitude compared to conventional patch-clamping. This kind of throughput, though still not within the realm needed for high-throughput random screening of compounds (see below), is highly suitable for targeted subset screening, secondary screening, high-throughput characterisation studies and to the support medicinal chemistry programmes.

Although this clearly represents a major advance, there are certain potential disadvantages of the IonWorks system. Due to the mechanics of the instrument, which

---

VL/VA, ventral thalamic nuclei; iml, internal medullary lamina of the thalamus; MDD, Mdfa, MDFi, MDV, medial dorsal thalamic nuclei; scale bar = 5 mm). B) Magnified views of basal ganglia/thalamus showing differential subcellular localisation of Na$_V$1.2 (scale bars = 100 μM). Na$_V$1.2 (left hand panels, II) shows axonal staining with immunopositive fibre tracts in the external global pallidus (GPe – arrows) and in the thalamus, passing into the internal capsule (Ic) from the reticular thalamic nucleus (rt – arrows). In contrast, Na$_V$1.3 and Na$_V$1.6 (panels III and VI) show immunostaining in the soma of cells in the reticular thalamic nucleus (arrows) and, for Na$_V$1.3, also in the associated processes (arrowheads). C) Higher magnification of the somato-motor cortex, (scale bar represents 100 μM) showing Na$_V$1.1, 1.3 and 1.6 staining in the soma (arrows), proximal processes (arrowheads) and axon hillocks (asterisks) of cortical neurons. (Reproduced from [82] with permission from Elsevier)

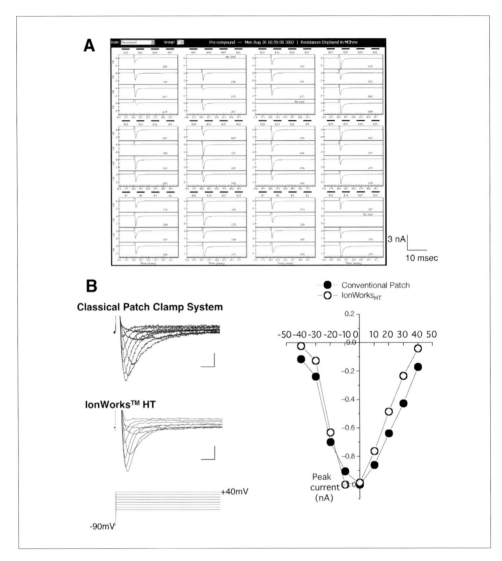

Figure 8

Human Na$_V$1.3 currents measured by IonWorks automated electrophysiology instrument
A) A screen dump from the IonWorks real-time display showing 48 simultaneous recordings
made using a stable CHO cell line expressing human Na$_V$1.3. B) Biophysical measurements
made by Ionworks are almost indistinguishable from those made with conventional patch-
clamp electrophysiology. Left: representative inward currents evoked by a series of depolar-
ising pulses. Right: current voltage relationship measured by conventional patch-clamp (•)
and Ionworks (○). (Reproduced from [90] with permission from the Society for Biomolecu-
lar Screening)

requires re-positioning of the recording electrode head during compound addition, it is not feasible to record some types of ligand-gated which have very fast kinetics. Also in comparison to conventional patch-clamp (and some of the other automated systems under development) the seal resistances obtained using the IonWorks patch plate are very low (e.g., 100 MΩ *versus* > 1 GΩ). This gives rise to greater ionic leak currents and potential errors in leak subtraction. Moreover, the inability to compensate for series resistance and capacitance could result in potential errors and variability in measurements of clamped voltage or current amplitude with consequent errors in estimation of drug effects. In practice, when measuring Na$_V$ currents of 2 nA or less these factors are not limitations, and the biophysical and pharmacological measurements obtained using IonWorks are almost indistinguishable to those made with conventional patch-clamp ([90], Fig. 8b). Crucially for the development of novel therapeutic Na$_V$ blockers, IonWorks can be used to measure use-dependent block (Fig. 9) providing assays that can be used to monitor this parameter during secondary screening, or to further optimise this during the development of hit compounds into lead series. Another consideration with IonWorks, and similar planar array instruments, is that they place great demands on the cell line being used for screening, since they are single cell assays in which the cell is randomly selected. Thus, a very high proportion of cells in the population that express usable currents is essential in order to avoid low success rates and unacceptably high costs resulting from compound wastage and reduced throughput. On the other hand, another important application of these instruments is during cell line generation. These instruments enable a vast increase in the number of clones that can be screened compared to that previously possible using conventional patch-clamp. Using IonWorks this approach has been highly successful for a variety of different ion channels, including Na$_V$s, and cell lines in which greater than 90–95% of cells express currents can be isolated (Fig. 10).

Although the advent of automated electrophysiology platforms represents a massive step forward for Na$_V$ and other ion channel drug development it is not yet feasible to apply these to primary high-throughput screening. This is largely due to the size of compound collections now being screened by most major pharmaceutical companies, which makes both the cost and the timeframe involved prohibitive, especially when compared to other high-throughput screening platforms. Considerable improvements in throughput, for example by increased parallelisation, and reduction in consumable costs (e.g., patch plates), are required to make this a more feasible option. In the meantime, alternative assay formats are required for the primary screening stages, and a number of these can be considered for assaying Na$_V$ channels. Ion flux assays, for example using Li$^+$ or radiolabeled guanidinium as tracer ions, have been advocated [91] though both have limited throughputs and the latter presents significant health and safety issues in an HTS environment. Fluorescent ion binding dyes are widely used for calcium channel assays, but there are no corresponding dyes that are sufficiently selective for Na$^+$ ions. This has led to wide-

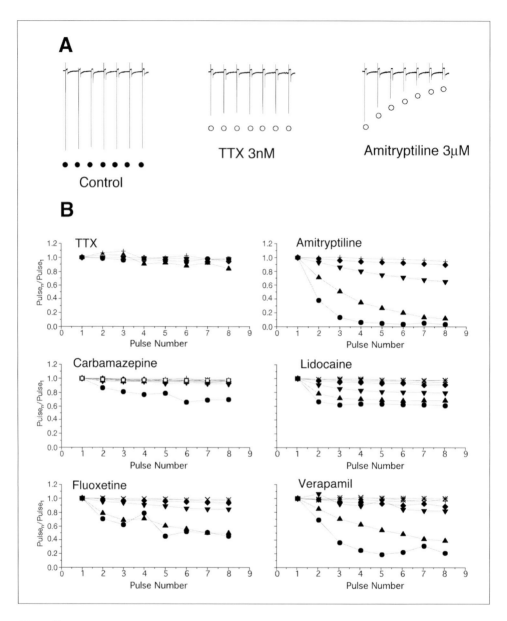

Figure 9

Measurement of use-dependent drug action using IonWorks automated electrophysiology
A) CHO cells stably expressing human Na$_V$1.3 were given a series of pulses (from a holding
potential of –90 mV) to 0 mV of 20 msec duration at a frequency of 9 Hz. In the absence of
drug there is no decay in peak current (control series). In the presence of tetrodotoxin (3 nM)
the first pulse is reduced by about 50% but there is no further decrease in subsequent puls-

spread use of fluorescent dyes that respond to changes in membrane potential [92], such as the lipophilic negatively charged oxanol dye, DiBAC$_4$(3). This has low fluorescence in an extracellular aqueous environment but upon cellular depolarisation it redistributes into the cytosol where it has increased emission due to interaction with cellular membranes. This can be used successfully for Na$_V$ channel assays and can be configured for HTS using a fluorescence imaging plate reader (FLIPR). Because the response of the dye is extremely slow (seconds) in relation to channel kinetics (milliseconds), the assay is dependent on the depolarisation event being sustained for long enough to obtain a measurable signal. For the brain Na$_V$ subtypes this can be addressed by using toxins such as α-scorpion toxin to prevent inactivation (Fig. 11). However, in some cell backgrounds the signal can be blunted or even ablated due to the presence of endogenous channels (e.g., voltage-gated potassium channels) or other mechanisms that may serve to rapidly restore the resting potential of the cell. To address this issue improved voltage sensitive dyes with faster kinetics have now been developed, for example the voltage sensitive probe (VSP) dyes developed by Aurora Biosciences [93]. This system depends on fluorescence resonance energy transfer (FRET) between two different dyes, which also provides a ratiometric measurement having the advantage of greater sensitivity as well as giving an internal control. The first dye is a coumarin-linked phospholipid (CC2-DMPE) that acts as the FRET donor and is fixed in the outer leaflet of the plasma membrane. The second is a mobile voltage-sensitive oxanol dye (DiSBAC$_2$(3)) that partitions within the plasma membrane. At relatively negative resting potentials the oxanol is distributed near to the outer membrane leaflet where it acts as a FRET acceptor when excited by the coumarin donor. Depolarisation results in rapid translocation of the oxanol to the inner surface of the membrane, with resultant decrease in oxanol (FRET) fluorescence and increase in coumarin (non-FRET) fluorescence. This system results in kinetics that are ~100-fold faster than DiBaC$_4$(3) assays [93] and works very effectively for Na$_V$ assays (Fig. 11c). Such assays can be

---

es. With the use-dependent blocker, amitryptiline (3 μM) peak current decreases with each pulse due to the use-dependent accumulation of channels inhibited. B) Concentration dependence of use-dependent block for a number of sodium channel drugs. For each drug a number of three-fold serial dilutions (indicated by ×, △, ♦, ▼, ▲, ● respectively) were tested using the pulse protocol described above. For each concentration of drug the peak current for each pulse is plotted as a ratio of the first pulse in the series. The highest concentrations used for each drug (indicated by +) were as follows: TTX (12 nM), amitryptilline (13 μM), carbemazepine (3.3 mM), lidocaine (3.3 mM), verapramil (111 μM), fluoxetine (37 μM). With the exception of tetrodotoxin, the extent of use dependent block increases with increasing concentration of drug used. (Reproduced from [90] with permission from Society for Biomolecular Screening)

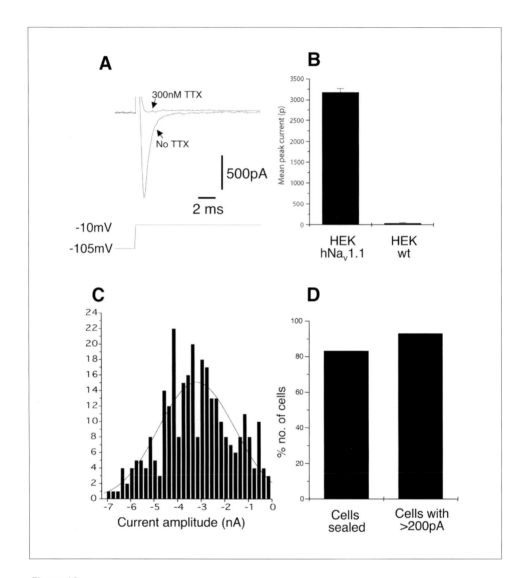

Figure 10

Use of IonWorks automated electrophysiology during cell line generation

HEK293 cells were transfected with an expression vector for human Na$_V$1.1 and multiple stable clones were selected, expanded and then screened for expression of Na$_V$ currents using IonWorks. The best expressing clone was identified and then characterised further. A) Representative inward Na$^+$ current evoked by depolarisation, showing blockade by tetrodotoxin (TTX, 300 nM). Mean peak currents (B) and distribution of peak current (C) measured from 270 cells are shown. D) Greater than 80% of cells gave usable seals (< 80 MΩ) and ~95% of these cells expressed currents above background.

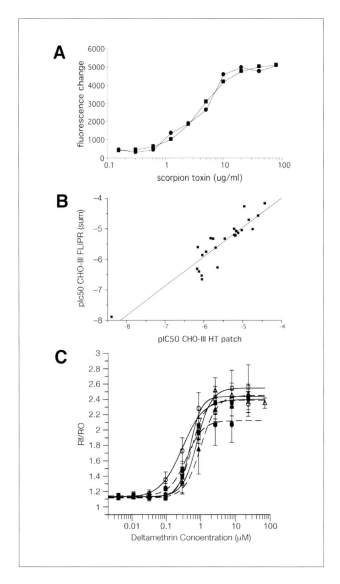

Figure 11
Membrane potential-based high-throughput screening assays for Na$_V$ channels
A) Concentration response curves for scorpion toxin venom mediated human Na$_V$1.3 activity measured in a FLIPR-DiBAC$_4$(3) assay – two representative curves are shown. B) Correlation of the potencies of a set of 23 different lamotrigine analogues and tetrodotoxin measured in the FLIPR-DiBAC$_4$(3) compared to patch-clamp electrophysiology. C) Concentration-response curves for deltamethrin-induced NaV1.8 activity measured in a VIPR-VSP assay.

configured for HTS and are capable of detecting Na_V inhibitors in high-throughput screening of large compound collections, though false positives can be a problem. Additionally, correlation between inhibitor potencies measured using membrane potential assays compared to patch-clamp is variable. For these reasons, once potential 'hit' compounds have been identified, filtering out of false positives and further characterisation requires secondary assays using electrophysiological techniques such as those described above.

Prospects for rational approaches to aid development of improved Na_V inhibitors

Advances in our understanding of Na_V channel structure–function, together with the advent of the first 3-D crystal structures for ion channels, provide potential opportunities to include structural insights in the development of new Na_V inhibitors. Although speculative at present, such knowledge-based approaches may become important aspects of Na_V channel drug discovery in the future. Three lines of research are actively being pursued; molecular characterisation of the binding site(s) of Na_V inhibitors, homology modelling of the Na_V channel pore, and pharmacophore modelling of known Na_V inhibitors.

Characterisation of the drug binding site(s) on brain Na_V channels has been an active area of investigation for several years. Attention initially focussed on the S6 transmembrane segment of domain IV (IVS6) since photo-affinity labelling had previously located a binding site for pore-blocking drugs in the corresponding region of voltage-gated calcium channels. Using alanine-scanning mutagenesis, two key amino acids, F1764 and Y1771, were identified in IVS6 of rat $Na_V1.2$ that were critical for high affinity binding of the local anaesthetic, etidocaine [94]. Remarkably, these two amino acids were also found have key interactions with a diverse range of other Na_V inhibitors, including antiarrhythmics (e.g., lidocaine, mexiletine, flecainide), anticonvulsants (e.g., phenytoin, lamotrigine) and anti-ischaemics (e.g., sipatrigine, crobenetine) [95–98]. The relative importance of these two residues varied for each compound tested. A similar analysis was subsequently carried out on the S6 segments from the other three domains. Three key amino acids for etidocaine binding were identified in IIIS6, L1465, N1466 and I1469, though only two of these were important for lamotrigine [99]. A single amino acid, I409, was identified in IS6 as being important for etidocaine binding, though this did not affect sipatrigine binding; no amino acids in IIS6 were found to be important for binding of either compound [100]. In summary, the picture that has emerged is that a diverse range of Na_V inhibitors share common elements within an overlapping binding site located in the inner pore of the channel. These common interactions vary in strength and additional compound-specific interactions also contribute to affinity. Importantly, the key amino acid binding determinants noted above are very highly conserved within the different Na$_V$ subtypes, though some differences can be seen in adjacent residues. Consistent with this, substitution of the equivalent key amino acids in IVS6 of the rat Na$_V$1.3 orthologue (F1710 and Y1717) gave similar results to that seen for Na$_V$1.2, reducing the affinity of the local anaesthetic, tetracaine [101].

The publication of the first experimentally determined 3-D structure for a voltage-gated ion channel [102], followed by the first structure of a channel in the open conformation [103], have given impetus to the structural modelling of Na$_V$ channels. These tetrameric bacterial K$^+$ channels both contain subunits with only two transmembrane segments (M1 and M2) flanking a pore loop (P region). However, this overall architecture is proposed to be analogous to the pore region of Na$_V$ channels, which consist of the SS1–SS2 pore loop flanked by the S5 and S6 transmembrane segments from each of the four domains. Thus, the bacterial channel structures provide a useful framework for molecular modelling of the Na$_V$ pore-region. A model for the closed state of Na$_V$1.4 has been generated by transferring the helical coordinates from KcsA after aligning the S6 and M2 segments using bulky aromatic residues and conserved glycines and constraining the known S6 local anaesthetic binding residues to face the inside of the pore [104]. The S5 segments were also positioned using conserved glycines by assuming these dock with bulky aromatic residues at the N-end of the S6 segments forming an "inverted tepee" in a similar manner to that found in M1 and M2 of KcsA. The pore loops of the two channels are expected to be dissimilar, since in Na$_V$ channels ion selectivity is determined by side-chain interactions of a single amino acid in each domain (DEKA) rather than carbonyls from four contiguous backbone amino acids per subunit as in KcsA. Thus, the Na$_V$ P loop was modelled separately as an α-helix-turn-β-strand motif and then docked into the inverted tepee in a manner compatible with interaction of the pore blocker, tetrodotoxin, with the residues which it is known to bind.

The structural model of the Na$_V$ pore described above is reasonably consistent with most of the available biophysical and mutational information available. However, for the purposes of analysing drug-binding this can be further refined by developing models that attempt to mimic the inactivated state, e.g., by docking the known solution structure of the inactivation gate [105] and by incorporating recent information from the structure of an open channel. The open state MthK structure also helps refine the alignment of transmembrane segments and underlines the importance of the conserved S6 glycine residues since they appear to play a key role in channel opening by acting as "hinges" that allow the helix to bend. Model structures incorporating these refinements for the Na$_V$1.8 subtype in the closed, open and inactivated states are shown in Figure 12a [Holger Scheib, Department of Structural Biology and Bioinformatics, University of Geneva and Swiss Institute of Bioinformatics, personal communication]. In these proposed structures the P region is separately modelled on an α-helix-turn-β-sheet template motif identified from a non-redundant subset of the PDB structure database and they also incorporate findings from more recent mutational studies which help position the IS6 and IIIS6 seg-

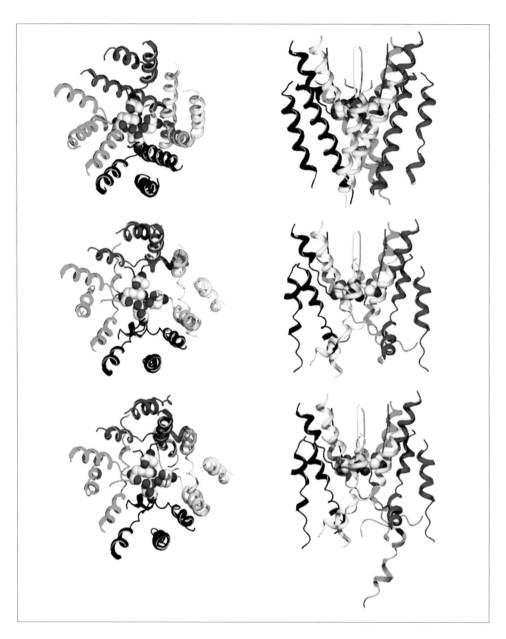

Figure 12
Homology model of the pore region of the Na$_V$1.8 channel and its interaction with lidocaine
A) Model structures of the closed (top), open (middle), and inactivated (bottom) forms of
the pore region in ribbon representation in top (left) and side (right) views. Ribbons are
coloured according to domain: domain I in yellow, domain II in blue, domain III in green,

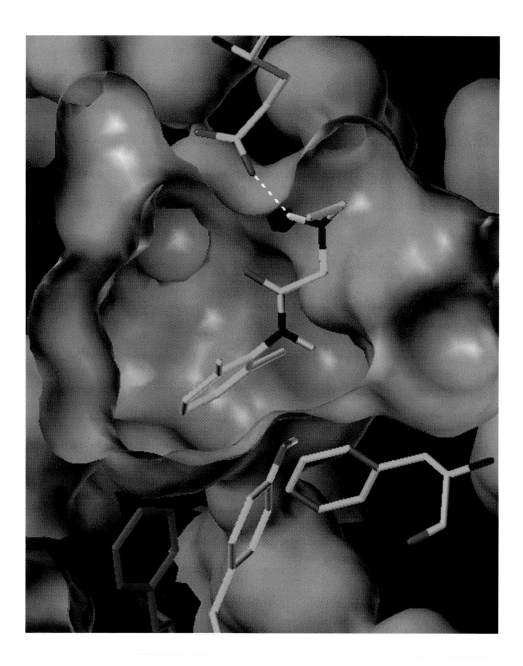

and domain IV in red. The inactivation gate substructure is shown in pink. The side chains of the DEKA sectivity filter residues are shown and coloured by CPK. B) Close-up side view of lidocaine docked into the fast-inactivated open form. The protein is represented by its Connolly surface (calculated with sybyl6.9) and is colored in orange.

ments [99, 101]. A close up view of a possible drug binding site in the inactivated state showing proposed interactions between lidocaine and key amino acid residues is shown in Figure 12b [Holger Scheib, personal communication]. The model also predicts additional interactions that can be tested experimentally.

A third approach that could potentially aid the rational development of novel Na_V blocking agents is the generation of structural models that attempt to describe the generic features of known inhibitors. While several groups have carried out structure–activity studies on related classes of compound (e.g., local anaesthetics [106], Diphenyl ureido-containing compounds [107], lamotrigine analogues [78]), there are relatively few reported attempts to identify a common pharmacophore from chemically diverse Na_V inhibitors. Given that all classes of Na_V drugs appear to use the same binding site, such pharmacophore models could be used predictively to design new molecules that can be tested experimentally, giving further data that can be used to refine the model. In this manner, Unverferth et al. [108] compared five well-known structurally different anticonvulsant compounds carbamazepine, phenytoin, lamotrigine, zonisamide and rufinamide. Common elements were recognised, in so much as each has at least one aryl ring (R), one electron donor atom (D) and a second electron donor atom in close proximity to the NH group forming a hydrogen bond donor/acceptor unit (HAD). A model was suggested incorporating these elements and the molecular distances between them which was validated and refined using another set of structurally diverse compounds (AWD 140-190, vinpocetine, dezinamide, remacemide). The model was subsequently tested by synthesising 3-amino, 4-amino and 5-aminopyrazoles [109, 110] and the activities of these compounds were found to fit the predictions of the suggested model. Although this kind model is rather crude, as more and more data become available, for example from the application of automated electrophysiology platforms to profile increasingly large numbers of compounds, so it may become possible to develop and test increasingly sophisticated pharmacophore models. For instance, given the scale that such high quality data can now be generated it may become feasible to generate pharmacophores models that could be used to predict features such as use-dependence.

Summary and prospects

Recent advances in methods for assaying brain Na_V channels have greatly increased their tractability for drug development. As a result, the way has been opened for the discovery of a new generation of Na_V inhibitors with improved potency and use-dependence. There is also growing evidence that selectivity over other types of channel, and indeed over individual Na_V subtypes, are more achievable hurdles than previously predicted. In parallel with this, the cloning and expression of the human Na_V channel family is enabling a better understanding of their properties as well as of their roles in normal physiology and in various disease settings. Armed with this increased knowledge, it is anticipated that better selectivity, potency and use-dependence will lead to improved therapeutic efficacy of Na$_V$ inhibitors, although a clearer understanding of the *in vivo* consequences of these refinements still awaits the emergence of this new generation of Na$_V$ inhibitors.

Acknowledgements

I would like to acknowledge my friends, colleagues and collaborators both at GlaxoSmithKline and elsewhere who, though too numerous to mention, have supported and contributed to Na$_V$ research at GSK. Particular thanks go to Steve Burbidge, Yuhua Chen, Tim Dale, Matt Hall, Del Trezise, Andy Powell, Holger Scheib, Ian McLay, Will Whitaker, and Xinmin Xie for direct contributions to the figures and for agreeing to their inclusion in this chapter. Thanks are also due to Charles Large, Iain McLay, Andy Powell and Del Trezise for helpful comments on the manuscript.

References

1   Faught E, Matsuo FU, Schachter S, Messenheimer J, Womble GP (2004) Long-term tolerability of lamotrigine: data from a 6-year continuation study. *Epilepsy and Behaviour* 5: 31–36

2   Catterall WA, Goldin AL, Waxman SG (2003) International Union of Pharmacology. XXXIX. Compendium of Voltage-Gated Ion Channels: Sodium Channels. *Pharmacol Rev* 55: 575–578

3   Hainsworth AH, Stefani A, Calabrese P, Smith T, Leach M (2000) Sipatrigine (BW619C89) is a neuroprotective agent and a sodium and calcium channel inhibitor. *CNS Drug Rev* 6: 111–134

4   Xie X, Hagan RM (1998) Cellular and molecular actions of lamotrigine: possible mechanisms of efficacy in bipolar disorder. *Neuropsychobiol* 38: 119–130

5   Noda M, Shimizu S, Tanabe T, Takai T, Kayano T, Ikeda T, Takahashi H, Nakayama H, Kanoaka Y, Minamino N et al (1984) Primary structure of *Electrophorus electricus* sodium channel deduced from cDNA sequence. *Nature* 312: 121–127

6   Segal M (2002) Sodium channels and epilepsy electrophysiology. In: G Bock, J Goode (eds): *Sodium channels and neuronal hyperexcitability*. Wiley Press, West Sussex, UK, 173–188

7   Feher O, Erdelyi L, Papp A (1988) The effect of pentylenetetrazole on the metacerebral neuron of Helix pomatia. *Gen Physiol Biophys* 7. 505–516

8   Zhai J, Wieland SJ, Sessler FM (1997) Chronic cocaine intoxification alters hippocampal sodium channel function. *Neurosci Letts* 229: 121–124

9   Tian LM, Otoom S, Alkhadi KA (1995) Endogenous bursting due to altered sodium channel function in rat hippocampal CA1 neurons. *Brain Res* 680: 164–172

10   Dorman DC, Beasley VR (1991) Neurotoxicology of pyrethrin and pyrethroid insecticides. *Vet Human Toxicol* 33: 23–43

11   Segal MM (1994) Endogenous bursts underlie seizure-like activity in solitary hippocampal neurons in microcultures. *J Neurophysiol* 72: 1874–1884

12   Segal MM, Zurakowski D, Douglas AF (1995) Late sodium channel openings underlie ictal epileptiform activity, and are preferentially diminished by the anticonvulsant phenytoin. *Soc Neurosci Abstr* 21: 777

13   Chao TI, Alzheimer C (1995) Effects of phenytoin on the persistent Na$^+$ current of mammalian CNS neurones. *Neuroreport* 6: 1778–1780

14   Taverna S, Sancini G, Mantegazza M, Franceschetti S, Avanzini G (1999) Inhibition of transient and persistent Na$^+$ current fractions by the new anticonvulsant topiramate. *J Pharmacol Exp Ther* 288: 960–968

15   Spadoni, F, Hainsworth AH, Mercuri NB, Caputi L, Martella G, Lavaroni F, Bernardi G, Stefani A (2002) Lamotrigine derivatives and riluzole inhibit INaP in cortical neurons. *Neuroreport* 13: 1167–1170

16   Lossin C, Wang DW, Rhodes TH, Vanoye CG, George AL (2002) Molecular basis of an inherited epilepsy. *Neuron* 34: 877–884

17   Kearney JA, Plummer NW, Smith MR, Kapur J, Cummins TR, Waxman SG, Goldin AL, Meisler MH (2001) A gain-of-function mutation in the sodium channel gene SCN2A results in seizures and behavioural abnormalities. *Neurosci* 102: 307–317

18   Wallace RH, Wang DW, Singh R, Scheffer IE, George AL, Phillips HA, Saar K, Reis A, Johnson EW, Sutherland GR et al (1998) Febrile seizures and general epilepsy associated with the Na$^+$ channel β1 subunit gene, SCN1B. *Nat Genet* 19: 366–370

19   Ceulemans BP, Claes LR, Lagae LG (2004) Clinical correlations of mutations in the SCN1A gene: from febrile seizures to severe myoclonic epilepsy in infancy. *Ped Neurol* 30: 236–243

20   Lossin C, Rhodes TH, Desai RR, Vanoye CG, Wang D, Carnicui S, Devinsky O, George AL (2003) Epilepsy-associated dysfunction in the voltage-gated neuronal sodium channel SCN1A. *J Neurosci* 23: 11289–11295

21   Suguwara T, Tsurubuchi Y, Fujiwara T, Mazaki-Miyazaki E, Nagata K, Montal M, Inoue Y, Yamakawa K (2003) Na$_V$1.1 channels with mutations of severe myoclonic epilepsy in infancy display attenuated currents. *Epilepsy Res* 54: 201–207

22   Suguwara T, Tsurubuschi Y, Agarwala KL, Ito M, Fukuma G, Mazaki-Miyazaki E, Nagafuji H, Noda M, Imoto K, Wada K et al (2001) A missense mutation of the Na$^+$ channel alpha II subunit gene Na(v)1.2 in a patient with febrile and afebrile seizures causes channel dysfunction. *Proc Natl Acad Sci USA* 98: 6384–6389

23   Berkowitz SF, Heron SE, Giordano L, Marini C, Guerrini R, Kaplan RE, Gambardelli A, Steinlein OK, Grinton BE, Dean JT et al (2004) *Ann Neurol* 55: 550–557

24   Kamiya K, Kaneda K, Sugawara T, Mazaki E, Okamura N, Montal M, Makita N, Tanaka M, Fukushima K, Fujiwara T et al (2004) A nonsense mutation of the sodium channel gene SCN2A in a patient with intractable epilepsy and mental decline. *J Neurosci* 24: 2690–2698

25 Audenaert D, Claes L, Ceulemens B, Logfren A, Van Broeckhoven C, De Jonghe P (2003) A deletion in SCN1B is associated with febrile seizures and early onset absence epilepsy. *Neurol* 61: 854–856

26 Meadows LS, Malhotra J, Loukas A, Thyagarajan V, Kazen-Gillespie KA, Koopman MC, Kreigler S, Isom LL, Ragsdale DS (2002) Functional and biochemical analysis of a sodium channel β1 subunit mutation responsible for generalised epilepsy with febrile seizures plus type 1. *J Neurosci* 22: 10699–10709

27 Bartolomei F, Gastaldi M, Massacrier A, Planells R, Nicolas S, Cau N (1997) Changes in the mRNAs encoding subtypes I, II and III sodium channel alpha subunits following kainate-induced seizures in rat brain. *J Neurocytol* 26: 667–678

28 Gastaldi M, Bartolomei F, Massacrier A, Planells R, Robaglia-Schlupp A, Cau P (1997) Increase in mRNAs encoding neonatal II and III sodium channel α-isoforms during kainate-induced seizures in adult rat hippocampus. *Mol Brain Res* 44: 179–190

29 Aronica E, Yankaya B, Troost D, Van Vliet EA, Lopes da Silva FH, Gorter J (2001) Induction of neonatal sodium channel II and III α-isoform mRNAs in neurons and microglia after status epilepticus in the rat hippocampus. *Eur J Neurosci* 13: 1261–1266

30 Ketelaars SO, Gorter JA, van Vliet EA, Lopes da Silva FH, Wadman WJ (2001) Sodium currents in isolated rat CA1 pyramidal and dentate granule neurones in the post-status epilepticus model of epilepsy. *Neurosci* 105: 109–120

31 Ellerkmann RK, Remy S, Chen J, Sochivko D, Elger CE, Urban BW, Becker A, Beck H (2003) Molecular and functional changes in voltage-dependent Na$^+$ channels following pilocarpine-induced status epilepticus in rat dentate granule cells. *Neurosci* 119: 323–333

32 Remy S, Urban B, Elger CE, Beck H (2003) Anticonvulsant pharmacology of voltage-gated Na$^+$ channels in hippocampal neurons of control and chronically epileptic rats. *Eur J Neurosci* 17: 2648–2658

33 Remy S, Gabriel S, Urban BW, Dietrich D, Lehmann TN, Elger CE, Heinemann U, Beck H (2003) A novel mechanism underlying drug resistance in chronic epilepsy. *Ann Neurol* 53: 469–479

34 Agrawal N, Alonso A, Ragsdale DS (2003) Increased persistent sodium currents in rat entorhinal cortex layer V neurons in a post-status epilepticus model of temporal lobe epilepsy. *Epilepsia* 44: 1601–1604

35 Grigorenko E, Glazier S, Bell W, Tytell M, Nosel E, Pons T, Deadwyler SA (1997) Changes in glutamate receptor subunit composition in hippocampus and cortex in patients with refractory epilepsy. *J Neurol Sci* 153: 35–45

36 Lombardo AJ, Kuzniecky R, Powers RE, Brown GB (1996) Altered brain sodium channel transcript levels in human epilepsy. *Mol Brain Res* 35: 84–90

37 Whitaker WRJ, Faull RLM, Emson PC, Clare JJ (2001) Changes in mRNAs encoding voltage-gated sodium channel types II and III in human epileptic hippocampus. *Neurosci* 106: 275–285

38 Vreugdenhil M, van Veelen CWM, van Rijen PC, Lopes da Silva FH, Wadman WJ

(1998) Effect of valproic acid on sodium currents in cortical neurons from patients with pharmaco-resistant temporal lobe epilepsy. *Epilepsy Res* 32: 309–320

39 Reckziegel G, Beck H, Schramm J, Elger CE, Urban BW (1998) Electrophysiological characterisation of Na$^+$ currents in acutely isolated human hippocampal dentate granule cells. *J Physiol* 509: 139–150

40 Vreugdenhil M, Hoogland G, van Veelen CWM, Wadman WJ (2004) Persistent sodium curent in subicular neurons isolated from patients with temporal lobe epilepsy. *Eur J Neurosci* 19: 2769–2778

41 Reckziegel G, Beck H, Schramm J, Urban BW, Elger CE (1999) Carbamazepine effects on Na$^+$ currents in human dentate granule cells from epileptogenic tissue. *Epilepsia* 40: 401–407

42 Quiroz JA, Singh J, Gould TD, Denicoff KD, Zarate CA, Manji HK (2004) Emerging experimental therapeutics for bipolar disorder: clues from the molecular pathophysiology. *Molec Psychiatry* 9: 756–776

43 Berns GS, Nemeroff CB (2003) The neurobiology of bipolar disorder. *Am J Med Genet* 123C: 76–84

44 Machado-Vieira R, Kapczinksy, Soares JC (2004) Perspectives for the development of animal models of bipolar disorder. *Prog Neuro Psy* 28: 209–224

45 Calabrese JR, Bowden CL, Sachs, Ascher JA, Monaghan ET, Rudd GD (1999) A double-blind placebo-controlled study of lamotrigine monotherapy in outpatients with bipolar I depression. *J Clin Psychiatry* 60: 79–88

46 Calabrese JR, Suppes T, Bowden CL, Sachs GS, Schwann AC, McElroy SL, Kusamakar V, Ascher JA, Earl NL, Greene PL et al (2000) A double-blind placebo-controlled prophylaxis study of lamotrigine in rapid cycling bipolar disorder J Clin Psychiatry 61: 841– 850

47 Post RM, Weiss SRB (1995) The neurobiology of treatment-resistant mood disorders. In: Bloom FE, Kupfer DJ (eds): *Psychopharmacology: The fourth generation of progress.* Raven Press, 1099–1111

48 Ketter TA, Manji HK, Post RM (2003) Potential mechanisms of action of lamotrigine in the treatment of bipolar disorders. *J Clin Psychopharmacol* 23: 484–495

49 Stratton SC, Large CH, Cox B, Davies G, Hagan RM (2003) Effects of lamotrigine and levetiracetam on seizure development in a rat amygdala kindling model. *Epilepsy Res* 53: 95–106

50 Drevets WC (2000) Neuroimaging abnormalities in the amygdala in mood disorders. *Ann NY Acad Sci* 985: 420–444

51 Peak H (1939) Time order error in successive judgements and in reflexes. I. Inhibition of the judgement and the reflex. *J Exp Psychol* 25: 535–565

52 Hoffmann HS, Searle (1965) Acoustic variables in the modification of startle reaction in the rat. *J Comp Physiol Psychol* 60: 53–58

53 Koch M, Robbins TW (2001) Special issue on the psychopharmacology of prepulse inhibition: basic and clinical studies. *Psychopharmacol* 156: 115–116

54   Braff DL, Grillon C, Geyer MA (1992) Gating and habituation of the startle reflex in schizophrenia patients. *Arch Gen Psychiatry* 49: 206–215

55   Braff DL, Geyer MA, Swerdlow NR (2001) Human studies of pre-pulse inhibition of startle: normal subjects, patients groups and pharmacological studies. *Psychopharmacol* 156: 234–258

56   Perry W, Minassian A Feifel D, Braff DL (2001) Sensorimotor gating deficits in bipolar disorder patients with acute psychotic mania. *Biol Psychiatry* 50: 418–424

57   Linn GS, Javitt DC (2001) Phencyclidine (PCP)-induced deficits of pre-pulse inhibition in monkeys. *Neuroreport* 12: 117–120

58   Mansbach RS, Geter MA (1989) Effects of phenylcyclidine and phenylcyclidine biologs on sensorimotor gating in the rat. *Neuropsychopharmocol* 2: 299–308

59   Brody SA, Geyer MA, Large CH (2003) Lamotrigine prevents ketamine but not amphetamine-induced deficits in prepulse inhibition in mice. *Psychopharmacol* 169: 240–246

60   Anand A, Charney DS, Oren DA, Berman RM, Hu XS, Cappiello A, Crystal JH (2000) Attenuation of the neuropsychiatric effects of ketamine with lamotrigine support for hyperglutamatergic effects of N-methyl-D-aspartate receptor antagonists. *Arch Gen Psychiatry* 57: 270–276

61   Calabrese P, Cupini LM, Centonze D, Pisani F, Bernardi G (2003) Antiepileptic drugs as a possible neuroprotective strategy in brain ischemia. *Ann Neurol* 53: 693–702

62   Taylor CP, Meldrum BS (1995) Na$^+$ channels as targets for neuroprotective drugs. *Trends In Pharm Sci* 16: 309–316

63   Kiedrowski L, Wroblewski JT, Costa E (1994) Intracellular sodium concentration in cultured cerebellar granule cells challenged with glutamate. Mol Pharmacol 45: 1050–1054

64   George AL, Knittle TJ, Tamkum MM (1992) Molecular cloning of an atypical voltage-gated sodium channel expressed in human heart and uterus: evidence for a distinct gene family. *Proc Natl Acad Sci USA* 89: 4893–4897

65   Watanabe E, Fujikawa A, Matsunaga H, Yasoshima Y, Sako N, Yamamoto T, Saegusa C, Noda M (2000) Na$_V$2/NaG channel is involved in control of salt-intake behaviour in the CNS. *J Neurosci* 20: 7743–7751

66   Kayano T, Noda M, Flockerzi V, Takahashi H, Numa S (1988) Primary structure of rat brain sodium channel III deduced from the cDNA sequence. *FEBS Letts* 228: 187–194

67   Noda M, Ikeda T, Kayano T, Suzuki H, Takeshima H, Kurasaki M, Takahashi H, Numa S (1986) Existence of distinct sodium channel messenger RNAs in rat brain. *Nature* 320: 188–192

68   Catterall WA (2000) From ionic currents to molecular mechanisms: the structure and function of voltage-gated sodium channels. *Neuron* 26: 13–25

69   Goldin AL (2001) Resurgence of sodium channel research. *Ann Rev Physiol* 63: 871–894

70   Smith MR, Smith RD, Plummer NW, Meisler MH, Goldin AL (1998) Functional analysis of the mouse SCN8A sodium channel. *J Neurosci* 18: 6093–6102

71   Raman IM, Sprunger LK, Meisler MH, Bean BP (1997) Altered subthreshold sodium

currents and disrupted currents and disrupted firing patterns in Purkinje neurons of Scn8a mutant mice. *Neuron* 19: 881–891

72 Xie XM, Dale TJ, John VH, Cater HL, Peakman TC, Clare JJ (2001) Electrophysiological and pharmacological properties of the human brain type IIA Na$^+$ channel expressed in a stable mammalian cell line. *Pflug Arch* 441: 425–433

73 Chen YH, Dale TJ, Romanos MA, Whitaker WRJ, Xie XM, Clare JJ (2000) Cloning, distribution and functional analysis of the type III sodium channel from human brain. *Eur J Neurosci* 12: 4281–4289

74 Burbidge SA, Dale TJ, Powell AJ, Whitaker WRJ, Xie XM, Romanos MA, Clare JJ (2002) Molecular cloning, distribution and functional analysis of the Na$_V$1.6 voltage-gated sodium channel from human brain. *Mol Brain Res* 103: 80–90

75 John VH, Main MJ, Powell AJ, Gladwell ZM, Hick C, Sidhu HS, Clare JJ, Tate S, Trezise DJ (2004) Heterologous expression and functional analysis of rat Na$_V$1.8 (SNS) voltage-gated sodium channels in the dorsal root ganglion neuroblastoma cell line ND7-23. *Neuropharmacol* 46: 425–438

76 Chen Y, Pennington NJ (2000) Competition between internal AlF$_4^-$ and receptor mediated stimulation of dorsal raphe neuron G-proteins coupled to calcium current inhibition. *J Neurophysiol* 83: 1273–1282

77 Ma JY, Catterall WA, Scheuer T (1997) Persistent sodium currents through brain sodium channels induced by G-protein βγ subunits. *Neuron* 19: 443–452

78 Clare JJ, Tate SN, Nobbs M, Romanos MA (2000) Voltage-gated sodium channels as therapeutic targets. *Drug Discovery Today* 5: 506–520

79 Faravelli L, Maj R, Veneroni O, Fariello RG, Benatti L, Salvati P (2000) NW-1029 is a novel Na$^+$ channel blocker with analgesic activity in animal models. *Soc Neurosci 30th Ann Meeting* 26: 1218

80 Oliveira JS, Redaelli E, Zaharenko AJ, Cassulini R, Konno K, Pimenta DC, De Freitas JC, Clare JJ, Wanke E (2004) Binding specificity of sea anemone toxins to Nav 1.1–1.6 sodium channels: Unexpected contributions from differences in the IV/S3-S4 outer loop. *J Biol Chem* 279: 33323–33335

81 Li RA, Ennis IL, Xue T, Nguyen HM, Tomaselli GF, Glodin AL, Marban E (2003) Molecular basis for isoform-specific micro-conotoxin block of cardiac, skeletal muscle and brain Na$^+$ channels. *J Biol Chem* 278: 8717–8724

82 Whitaker WRJ, Faull RLM, Waldvogel H, Plumpton CJ, Emson PC, Clare JJ (2001) Comparative distribution of voltage-gated sodium channel proteins in human brain. *Mol Brain Res* 88: 37–53

83 Whitaker WRJ, Clare JJ, Powell AJ, Chen Y, Faull RLM, Emson PC (2000) Distribution of voltage-gated sodium channel α and β subunit mRNAs in human cerebellum, cortex and hippocampal formation. *J Comp Neurol* 422: 123–139

84 Coward K, Plumpton C, Facer P, Birch R, Carlstedt T, Tate S, Bountra C, Anand P (2000) Immunolocalisation of SNS/PN3 and NaN/ SNS2 sodium channels in human pain states. *Pain* 85: 41–50

85 Coward K, Aitken A, Powell A, Plumpton C, Birch R, Tate S, Bountra C, Anand P

(2001) Plasticity of TTX-sensitive sodium channels PN1 and brain III in injured human nerves. *Neuroreport* 12: 495–499

86 Trimmer JS, Rhodes KJ (2004) Localisation of voltage-gated ion channels in mammalian brain. *Ann Rev Physiol* 66: 477–519

87 Lindia JA, Abbadie C (2003) Distribution of the voltage-gated sodium channel Na$_V$1.3-like immunoreactivity in the adult rat central nervous system. *Brain Res* 960: 132–141

88 Caldwell JH, Schaller KL, Lasher RS, Peles E, Levinson SR (2000) Sodium channel Na$_V$1.6 is localised at nodes of Ranvier, dendrites and synapses. *Proc Natl Acad Sci USA* 97: 5616–5620

89 Wood C, Williams C, Waldron GJ (2004) Patch-clamping by numbers. *Drug Discovery Today* 9: 434–441

90 Schroeder K, Neagle B, Trezise DJ, Worley J (2003) IonWorks: A new high-throughput electrophysiology measurement platform. *J Biomol Screening* 8: 50–64

91 Gill S, Gill R, Lee SS, Hesketh JC, Fedida D, Rezazadeh S, Stankowitch, Liang D (2003) Flux assays in high throughput screening of ion channels in drug discovery. *Assay Drug Technologies* 1: 709–717

92 Wolff C, Fuks B, Chatelan P (2003) Comparative study of membrane potential-sensitive fluorescent probes and their use in ion channel screening. *J Biomol Screening* 8: 533–543

93 Gonzalez JE, Oades K, Leychkis Y, Harootunian A, Negalescu PA (1999) Cell-based assays and instrumentation for screening ion channel targets. *Drug Discovery Today* 4: 431–439

94 Ragsdale DS, McPhee JC, Scheuer T, Catterall WA (1994) Molecular determinants of state-dependent block of Na$^+$ channels by local anesthetics. *Science* 265: 1724–1728

95 Ragsdale DS, McPhee JC, Scheuer T Catterall WA (1996) Common molecular determinants of local anesthetic, antiarryhthmic and anticonvulsant block of voltage-gated Na$^+$ channels. *Proc Natl Acad Sci USA* 93: 9270–9275

96 Weiser T, Qu Y, Catterall WA, Scheuer T (1999) Differential interaction of R-mexiletene with the local anesthetic receptor site on brain and heart sodium channels α-subunits. *Mol Pharmacol* 56: 1238–1244

97 Carter AJ, Grauert M, Pschorn U, Bechtel WD, Bartmann-Lindholm C, Qu Y, Scheuer T, Catterall WA, Weiser T (2000) Potent blockade of sodium channels and protection of brain tissue from ischemia by BIII 890 CL. *Proc Natl Acad Sci USA* 97: 4944–4949

98 Liu G, Yarov-Yarovoy V, Nobbs M, Clare JJ, Scheuer T, Catterall WA (2003) Differential interactions of lamotrigine and related drugs with transmembrane segment IVS6 of voltage-gated sodium channels. *Neuropharmacol* 44: 413–422

99 Yarov-Yarovoy V, Brown J, Sharp E, Clare JJ, Scheuer T, Catterall WA (2001) Molecular determinants of voltage-dependant gating and binding of pore-blocking drugs in transmembrane segment IIIS6 of the sodium channel α subunit. *J Biol Chem* 276: 20–27

100 Yarov-Yarovoy V, McPhee JC, Idsvoog D, Pate C, Scheuer T, Catterall WA (2002) Role of amino acid residues in transmembrane segments IS6 and IIS6 of the Na$^+$ channel a subunit in voltage-dependent gating and drug block. *J Biol Chem* 38: 35393–35401

101 Li HL, Galue A, Meadows L, Ragsdale DS (1999) A molecular basis for the different local anesthetic affinities of resting *versus* open and inactivated states of the sodium channel. *Mol Pharmacol 55*: 134–141

102 Doyle DA, Morais Cabral J, Pfuetzner RA, Kuo A, Gulbis JM, Cohen SL, Chait BT, MacKinnon R (1998) The structure of the potassium channel: molecular basis of K^+ conduction and selectivity. *Science* 280: 69–77

103 Jiang Y, Lee A, Chen J, Cadene M, Chait BT, MacKinnon R (2002) Crystal structure and mechanism of a calcium-gated potassium channel. *Nature* 417: 515–522

104 Lipkind GM, Fozzard HA (2000) KcsA crystal structure as framework for a molecular model of the $Na(^+)$ channel pore. *Biochemistry* 39: 8161–8170

105 Rohl CA, Boeckman FA, Baker C, Scheuer T, Catterall WA, Klevit RE (1999) Solution structure of the sodium channel inactivation gate. *Biochemistry* 38: 855–861

106 De Luca A, Natuzzi, Desaphy JL, Loni G, Lentini G, Franchini C, Tortorelli V, Camerino DC (2000) Molecular determinants of mexiletine structure for potent and use-dependent block of skeletal muscle sodium channels. *Mol Pharmacol* 57: 268–277

107 Snell LD, Claffey DJ, Ruth JA, Valenzuela F, Cardoso R, Wang Z, Levinson SR, Sather WA, Williamson AV, Ingersoll NC et al (2000) Novel structures having antagonist actions at both the glycine site of the N-methyl-D-aspartate receptor and neuronal voltage-sensitive sodium channels: biochemical electrophysiological and behavioural characterisation. *J Pharmacol Exp Ther* 292: 215–227

108 Unverferth K, Engel J, Hofgen N, Rostock A, Gunther R, Lankau HJ, Menzer M, Rolfs A, Liebscher J, Muller B et al (1998) Synthesis, anticonvulsant activity, and structure-activity relationships of sodium channel blocking 3-aminopyrroles. *J Med Chem* 41: 63–73

109 Lankau HJ, Menzer M, Rostock A, Arnold T, Rundfeldt C, Unverferth K (1999) 3-Amino- and 5-aminopyrazoles with anticonvulsant activity. *Arch Pharm (Weinheim)* 332: 219–221

110 Lankau HJ, Menzer M, Rostock A, Arnold T, Rundfeldt C, Unverferth K (1999) Synthesis and anticonvulsant activity of new 4-aminopyrazoles and 5-aminopyrazol-3-ones. *Pharmazie* 54: 705–706

111 Xie X, Lancaster B, Peakman T, Garthwaite J (1995) Interaction of the antiepileptic drug lamotrigine with recombinant rat brain type IIA Na^+ channels and with native Na+ channels in rat hippocampal neurones. *Pflugers Arch* 430: 437–446

112 Mantegazza M, Yu FH, Powell AJ, Clare JJ, Cetterall WA, Scheuer T (2005) Molecular determinants for modulation of persistent sodium current by G-protein βγ subunits. *J Neurosci* 25: 3341–3349

Voltage-gated sodium channels and visceral pain

Jennifer M.A Laird[1,3] and Fernando Cervero[2]

[1]Bioscience Department, AstraZeneca R&D Montréal, 7171 Frédérick-Banting, Ville Saint-Laurent, Quebec H4S 1Z9, Canada; [2]Anaesthesia Research Unit and Centre for Research on Pain, McGill University, Montréal, Quebec, Canada; [3]Pharmacology and Experimental Therapeutics, McGill University, Montréal, Quebec, Canada

Introduction

Pain is a highly dynamic process. An injury to the skin or to an internal organ sets in motion a chain of events leading to the perception of acute pain, to the generation of hypersensitive areas around the injury site (primary hyperalgesia) as well as remote from the lesion (secondary and referred hyperalgesia) and eventually to the establishment of a chronic pain state [1]. The nature of the originating lesion, the process of sensitization of the sensory receptors at the site of injury and the plastic changes of the central nociceptive pathways will determine the time course and the magnitude of the pain state. The most significant advances in pain research in the last few years have been the recognition of the dynamic nature of the pain pathway and the identification of the molecular elements responsible for the functional changes that lead to chronic pain and hyperalgesia (e.g., [2]).

Pain from internal organs – visceral pain – is the most common form of pain and afflicts virtually every human being at one time or another. Unlike somatic pain – pain from skin, muscle and joints, visceral pain is often dull, badly localized and difficult to describe [3]. The dynamic and changing nature of pain perception is perhaps most remarkable in the visceral pain domain. A particularly intriguing form of visceral pain is that known as 'functional' pain. Functional visceral pain is pain that occurs in the absence of demonstrable pathology of the internal organs or of its associated nerves. This is particularly well studied in the gastrointestinal (GI) tract. Patients with functional abdominal pain, for example, complain of discomfort, bloating or pain but after extensive clinical investigations nothing is found in the GI tract that could explain the sensory symptoms. Functional abdominal pain is the central symptom of irritable bowel syndrome (IBS), a condition characterized by discomfort, pain and alterations of defecation in the absence of peripheral pathology [3, 4]. Likewise, functional visceral pain characterises syndromes such as chronic pelvic pain, chronic prostatitis and interstitial cystitis. Functional visceral pain is commonly interpreted as a consequence of hypersensitivity of visceral nociceptive pathways, either of the sensory receptors in the periphery or of the central neurons [3, 5]. In this case the sensitization of the nociceptive pathway would be the mechanism for the enhanced pain perception even though the sensitizing process would not involve a demonstrable lesion in the originating organ [6, 7].

This process of enhanced sensitivity or sensitization, either peripheral or central, is therefore at the heart of all current interpretations of the pathophysiology of visceral pain. In the case of organic pain, peripheral sensitizing agents include inflammatory mediators and cytokines released at the injury site. In addition there are contributions from neuromodulators released by the sensory endings activated by the noxious stimuli, a process known as neurogenic inflammation. Central sensitization of central nervous system (CNS) neurons is triggered and maintained by the enhanced activity of the sensory afferents and amplified by the properties of the neural network [2]. The same mechanisms, peripheral and/or central have been proposed to apply in the case of functional visceral pain with the proviso that there should be no peripheral trigger to the process [7].

The sensory innervation of the viscera not only has a role in pain perception but also participates in the regulation and control of motility and secretion [6]. Therefore, any alteration in the excitability of sensory afferents will have a direct influence on the regulatory functions of the organ. Often, clinical symptoms associated with visceral lesions are the consequence of the hypersecretion or hypermotility caused by sensitized afferents. Alternatively, the mediators released at the periphery by inflammation, either neurogenic or non-neurogenic, can change the properties of the secretory and motor cells that in turn will affect the sensory signals arising from the inflamed area. Therefore it is almost impossible to separate the sensory alterations due to peripheral visceral lesions from the motor and secretory disturbances also caused by the lesion [8].

The main focus of this chapter is the role of voltage-gated sodium channels in the triggering and maintenance of sensitization of visceral sensory afferents. Voltage-gated sodium channels are essential for the propagation of action potentials along axons and also contribute to controlling membrane excitability. There are several sub-types of voltage-gated sodium channels expressed in primary sensory neurones. The sodium currents that they mediate are classified electrophysiologically into several types on the basis of their kinetics and their sensitivity to a natural toxin, tetrodotoxin (TTX) [9]. Almost all spinal ganglion neurones express TTX-sensitive sodium currents, but TTX-resistant currents seem to be associated preferentially with nociceptive primary afferent neurons [10, 11]. Modulation of the TTX-resistant sodium current has been proposed as a molecular substrate for the sensitization of nociceptors and regulation in the expression of the different sub-types of voltage-gated sodium channels is also thought to contribute to the enhanced excitability that characterises sensitization (see chapters by L.V. Dekker/ D. Cronk and M.S. Gold).

Expression of sodium channel sub-types in visceral afferent neurones

In somatic nerves, primary afferents identified as nociceptors are more likely than non-nociceptors to express the TTX-resistant channels sodium channel α subunits $Na_V1.8$ and $Na_V1.9$ [10, 11]. Similarly, work using isolated spinal ganglion neuron cell bodies has shown that TTX-resistant currents tend to be present in neurons that show other properties associated with nociceptors such as responses to inflammatory mediators or to capsaicin, the active component of chilli peppers [12–14].

Several groups have recently characterized the sodium channel currents in identified dorsal root ganglion somata innervating the viscera. These experiments rely on injecting a tracer into the gut wall, waiting for retrograde transport of the tracer to the cell body in the dorsal root ganglion (DRG) and then isolating the cells and recording from the labelled neurones. Using these methods, TTX-resistant currents have been found in the spinal afferent neurons innervating the stomach [15, 16], the ileum [17], the colon [18–20] and the bladder [21, 22].

The biophysical characteristics of the sodium currents present in colon afferents have been studied in greater detail [19, 20]. Almost all afferents (95–100%) tested showed a high-threshold, slowly-inactivating TTX-resistant current of the type produced by the α subunit of a TTX-resistant channel encoded by the $Na_V1.8$ gene [23]. Very few (0–12%) showed evidence of a persistent TTX-resistant current of the type encoded by $Na_V1.9$ [24]. This correlates with an immunohistochemical study of bladder afferent neurones in L6/S1 DRG showing that ~60% of them expressed NaV1.8 immunoreactivity, whereas only 1% of them expressed $Na_V1.9$ [25]. In contrast, $Na_V1.9$ was expressed in ~70% of non-bladder afferent neurones in the L6/S1 dorsal root ganglia [25].

TTX-sensitive currents are also expressed in visceral afferent neurones [15–22]. In neurones innervating the mouse colon, the biophysical properties of the TTX-sensitive current fit well with those described for the $Na_V1.7$ channel [20]. This is consistent with TTX-sensitive currents reported in non-selected DRG neurones, suggesting that there are no important differences in the TTX-sensitive subunits expressed in visceral neurones compared to neurones innervating other targets.

Contribution of voltage-gated sodium channels to the sensitization of visceral afferent neurones

Modulation of the TTX-resistant current has been proposed as a possible molecular substrate for sensitization of primary afferent neurones. Studies using isolated DRG neurones as model systems for the terminal endings of nociceptive afferents have shown that proinflammatory mediators like prostaglandin E_2 (PGE_2) and serotonin enhance the TTX-resistant current by a mechanism involving phosphorylation [12–14]. The TTX-resistant component of Na^+ currents is likely to be involved in spike initiation, so increasing the TTX-resistant current would be expected to increase the firing probability of afferents. TTX-resistant currents in identified colon DRG neurones are also increased by PGE_2 treatment [19], suggesting this is also a mechanism that could underlie sensitization of GI nociceptors.

Inflammation of visceral tissue may therefore modulate sodium currents in visceral afferent neurones and thereby produce sensitization of primary afferents. Recording intracellularly from the terminal endings is not possible with the methods available. However, experiments in DRGs isolated from animals with inflammation of the innervation target show that the expression of sodium currents in the soma is influenced by peripheral inflammation. Thus ileitis induced by instillation of trinitrobenzene sulfonic acid (TNBS) and cyclophosphamide cystitis produce an increase in excitability of isolated DRG neurones labelled from the inflamed viscus [17, 21]. This was manifested as a decrease in the threshold for action potential firing and an increase in the rate of depolarisation of TTX-resistant action potentials, suggesting an increase in TTX-resistant Na^+ currents [17, 21].

A similar increase in excitability is seen in neurones innervating the stomach after either a mild gastritis induced by administering iodoacetamide in drinking water or by producing gastric ulcers with acetic acid injections into the stomach wall [15, 16]. In these experiments an increase in the peak TTX-resistant current was observed and there was also a decrease in the threshold for activation of this current. Likewise, the sodium channel density increased in neurones labelled from the mouse colon after induction of colitis with TNBS [20]. A more detailed characterization of the sodium currents revealed that this increase was due to a 62% increase in the slow TTX-resistant current (likely mediated by $Na_V1.8$), which also showed a decrease in the activation threshold. There was no significant change in the fast TTX-sensitive currents or in the persistent TTX-resistant current [20].

An increase in TTX-resistant current density correlated with increased expression of NaV1.8 mRNA has also been observed in somatic DRG neurones after hind limb inflammation with carrageenan [26]. Therefore the mechanisms of sensitization of visceral afferents may include acute modulation of the TTX-resistant currents induced by inflammatory mediators acting on the afferent terminals and also in the longer term, an increased expression of the TTX-resistant sodium channel subunits, resulting in a greater TTX-resistant current density.

Functional role of voltage-gated sodium channels in visceral pain

Further evidence for an important role for TTX-resistant sodium channels in visceral nociceptor sensitization comes from experiments in mice with a null mutation in the gene encoding for the $Na_V1.8$ sodium channel subunit [27]. $Na_V1.8$ is exclusively expressed in primary sensory neurones [23], thus any change in visceral pain sensation in these mice is due to changes in the extrinsic afferents. Visceral pain and referred hyperalgesia in $Na_V1.8$ null mice and their wild type littermates were compared in tests which differed in the degree to which behaviour depends on spontaneous, ongoing firing in sensitized nociceptors [28].

Intracolonic isotonic saline, which produces a brief distension, and intraperitoneal acetylcholine are acute noxious stimuli that do not provoke sensitization of nociceptors, or evidence of referred hyperalgesia. $Na_V1.8$ null mice responded normally to these stimuli [28]. However, $Na_V1.8$ null mutants did show markedly reduced pain responses and no referred hyperalgesia to intracolonic capsaicin, a model in which pain behaviour is sustained by ongoing activity in nociceptors sensitized by the initial capsaicin application [29]. $Na_V1.8$ knockout mice also showed blunted pain and hyperalgesia to intracolonic mustard oil [28], which sensitizes nociceptors and also provokes tissue-damage, providing an ongoing stimulus [29]. The null mutants showed identical inflammatory responses compared to wild-type mice, so the differences in pain responses are unlikely to be secondary to an impairment of inflammation.

In contrast, $Na_V1.8$ null mice showed no differences from wild-type mice in the pain or referred hyperalgesia induced by cyclophosphamide cystitis [28]. Cyclophosphamide produces cystitis by gradual accumulation of toxic metabolites in the bladder, and thus is a model of tonic noxious chemical stimulation [30]. What accounts for the differential response of the $Na_V1.8$ null mutants to these different visceral stimuli? One possibility is that $Na_V1.8$ is expressed in colon but not bladder afferents. However, the majority of both bladder and colon afferent neurons express TTX-resistant currents [18, 22]. Thus it seems likely that the difference in behaviour is due to the sensitizing nature of the stimulus. The $Na_V1.8$ subunit appears to be essential for the expression of visceral pain behaviour generated by sensitization of visceral nociceptors, but not for either acute visceral pain responses or pain generated by a sustained tonic noxious input [28].

An involvement of $Na_V1.8$ in spontaneous firing in sensitized visceral nociceptors is also supported by the observations of Yoshimura and colleagues [31] in experiments examining the effects of knocking down $Na_V1.8$ expression using antisense methods. They used a stimulus that acutely sensitizes bladder afferents, infusion of dilute acetic acid into the bladder. They found that treatment with antisense oligonucleotides inhibited the expression of spinal Fos (a marker of neuronal activity) and abolished the bladder hyperreflexia induced after acetic acid infusion in animals treated with mismatch oligonucleotides [31].

Summary

In conclusion the TTX-resistant sodium current, especially that produced by the $Na_V1.8$ subunit, appears to be a strong candidate for a molecular substrate underlying sensitization of visceral afferent nociceptive neurons.

The visceral anti-nociceptive effects of agents that block sodium currents confirm an important role for these channels in visceral sensation. Intravenous lidocaine, a use-dependent sodium channel blocker, is effective in inhibiting both pseudoaffective reflex responses and spinal neuronal discharges to noxious distension of the colon [32]. Likewise, the sodium channel blockers mexiletine and carbamazepine dose-dependently inhibit the responses of nociceptive colonic afferent fibres to colorectal distension [33]. There have been very few clinical reports of the effects of sodium channel blockers on visceral pain [32, 34] although one report describes that systemic local anaesthetics were effective in relieving pain from the spleen [35]. However, indirect evidence comes from the observation that tricyclic antidepressant drugs like amitriptyline are regularly prescribed for functional visceral pain. Although these compounds likely exert their antidepressant effects by blocking the re-uptake of monoamines, many are also potent sodium channel blockers, and this feature may contribute to their effectiveness in some visceral pain patients.

References

1   Cervero F, Laird JM (1996) Mechanisms of touch-evoked pain (allodynia): a new model. *Pain* 68: 13–23

2   Hunt SP, Mantyh PW (2001) The molecular dynamics of pain control. *Nat Rev Neurosci* 2: 83–91

3   Cervero F, Laird JMA (1999) Visceral pain. *Lancet* 353: 2145–2148

4   Drossman DA, Camilleri M, Mayer EA, Whitehead WE (2002) AGA technical review on irritable bowel syndrome. *Gastroenterology* 123: 2108–2131

5   Lin C, Al Chaer ED (2003) Long-term sensitization of primary afferents in adult rats exposed to neonatal colon pain. *Brain Res* 971: 73–82

6   Cervero F (1994) Sensory innervation of the viscera: Peripheral basis of visceral pain. *Physiol Rev* 74: 95–138

7   Mayer EA, Gebhart GF (1994) Basic and clinical aspects of visceral hyperalgesia. *Gastroenterol* 107: 271–293

8   Laird JMA, Roza C, Cervero F (1997) Effects of artificial calculosis on rat ureter motility: peripheral contribution to the pain of ureteric colic. *Am J Physiol* 272: R1409–1416

9   Wood JN, Baker M (2001) Voltage-gated sodium channels. *Curr Opin Pharmacol* 1: 17–21

10   Djouhri L, Fang X, Okuse K, Wood JN, Berry CM, Lawson SN (2003) The TTX-resistant sodium channel $Na_V1.8$ (SNS/PN3): expression and correlation with membrane properties in rat nociceptive primary afferent neurons. *J Physiol* 550: 739–752

11   Fang X, Djouhri L, Black JA, Dib-Hajj SD, Waxman SG, Lawson SN (2002) The presence and role of the tetrodotoxin-resistant sodium channel $Na_V1.9$ (NaN) in nociceptive primary afferent neurons. *J Neurosci* 22: 7425–7433

12   Gold MS, Reichling DB, Shuster MJ, Levine JD (1996) Hyperalgesic agents increase a

tetrodotoxin-resistant Na$^+$ current in nociceptors. *Proc Natl Acad Sci USA* 93: 1108–1112

13   England S, Bevan S, Docherty RJ (1996) PGE$_2$ modulates the tetrodotoxin-resistant sodium current in neonatal rat dorsal root ganglion neurones *via* the cyclic AMP-protein kinase A cascade. *J Physiol* 495: 429–440

14   Cardenas CG, Del Mar LP, Cooper BY, Scroggs RS (1997) 5HT4 receptors couple positively to tetrodotoxin-insensitive sodium channels in a subpopulation of capsaicin-sensitive rat sensory neurons. *J Neurosci* 17: 7181–7189

15   Bielefeldt K, Ozaki N, Gebhart GF (2002) Experimental ulcers alter voltage-sensitive sodium currents in rat gastric sensory neurons. *Gastroenterol* 122: 394–405

16   Bielefeldt K, Ozaki N, Gebhart GF (2002) Mild gastritis alters voltage-sensitive sodium currents in gastric sensory neurons in rats. *Gastroenterol* 122: 752–761

17   Moore BA, Stewart TM, Hill C, Vanner SJ (2002) TNBS ileitis evokes hyperexcitability and changes in ionic membrane properties of nociceptive DRG neurons. *Am J Physiol Gastrointest Liver Physiol* 282: G1045–G1051

18   Su X, Wachtel RE, Gebhart GF (1999) Capsaicin sensitivity and voltage-gated sodium currents in colon sensory neurons from rat dorsal root ganglia. *Am J Physiol* 277: G1180–G1188

19   Gold MS, Zhang L, Wrigley DL, Traub RJ (2002) Prostaglandin E(2) modulates TTX-R I(Na) in rat colonic sensory neurons. *J Neurophysiol* 88: 1512–1522

20   Beyak MJ, Ramji N, Krol KM, Kawaja MD, Vanner SJ (2004) Two TTX-resistant Na$^+$ currents in mouse colonic dorsal root ganglia neurons and their role in colitis-induced hyperexcitability. *Am J Physiol Gastrointest Liver Physiol* 287: G845–855

21   Yoshimura N, De Groat WC (1999) Increased excitability of afferent neurons innervating rat urinary bladder after chronic bladder inflammation. *J Neurosci* 19: 4644–4653

22   Yoshimura N, White G, Weight F, De Groat WC (1996) Different types of Na$^+$ and K$^+$ currents in rat dorsal root ganglion neurones innervating the urinary bladder. *J Physiol* 494: 1–16

23   Akopian AN, Sivilotti L, Wood JN (1996) A tetrodotoxin-resistant voltage-gated sodium channel expressed by sensory neurons. *Nature* 379: 257–262

24   Cummins TR, Dib-Hajj SD, Black JA, Akopian AN, Wood JN, Waxman SG (1999) A novel persistent tetrodotoxin-resistant sodium current in SNS-null and wild-type small primary sensory neurons. *J Neurosci* 19: RC43

25   Black JA, Cummins TR, Yoshimura N, de Groat WC, Waxman SG (2003) Tetrodotoxin-resistant sodium channels Na$_V$1.8/SNS and Na$_V$1.9/NaN in afferent neurons innervating urinary bladder in control and spinal cord injured rats. *Brain Res* 963: 132–138

26   Tanaka M, Cummins TR, Ishikawa K, Dib-Hajj SD, Black JA, Waxman SG (1998) SNS Na$^+$ channel expression increases in dorsal root ganglion neurons in the carrageenan inflammatory pain model. *Neuroreport* 9: 967–972

27   Akopian AN, Souslova V, England S, Okuse K, Ogata N, Ure J, Smith A, Kerr BJ, McMahon SB, Boyce S et al (1999) The tetrodotoxin-resistant sodium channel SNS has a specialized function in pain pathways. *Nat Neurosci* 2: 541–548

28 Laird JMA, Souslova V, Wood JN, Cervero F (2002) Deficits in visceral pain and referred hyperalgesia in Na$_V$1.8 (SNS/PN3)-null mice. *J Neurosci* 22: 8352–8356

29 Laird JMA, Martinez-Caro L, Garcia-Nicas E, Cervero F (2001) A new model of visceral pain and referred hyperalgesia in the mouse. *Pain* 92: 335–342

30 Olivar T, Laird JMA (1999) Cyclophosphamide cystitis in mice: behavioural characterization and correlation with bladder inflammation. *Eur J Pain* 3: 141–149

31 Yoshimura N, Seki S, Novakovic SD, Tzoumaka E, Erickson VL, Erickson KA, Chancellor MB, De Groat WC (2001) The involvement of the tetrodotoxin-resistant sodium channel Na$_V$1.8 (PN3/SNS) in a rat model of visceral pain. *J Neurosci* 21: 8690–8696

32 Ness TJ (2000) Intravenous lidocaine inhibits visceral nociceptive reflexes and spinal neurons in the rat. *Anesthesiol* 92: 1685–1691

33 Su X, Joshi SK, Kardos S, Gebhart GF (2002) Sodium channel blocking actions of the kappa-opioid receptor agonist U-50,488 contribute to its visceral antinociceptive effects. *J Neurophysiol* 87: 1271–1279

34 Glazer S, Portenoy RK (1991) Systemic local anesthetics in pain control. *J Pain Sympt Manage* 6: 30–39

35 Parris WC, Gerlock AJ Jr, MacDonell RC Jr (1981) Intra-arterial chloroprocaine for the control of pain associated with partial splenic embolization. *Anesth Analg* 60: 112–115

The functional interaction of accessory proteins and voltage-gated sodium channels

Kenji Okuse[1,2] and Mark D. Baker[3]

[1]Wolfson Institute for Biomedical Research, University College London, Gower Street, London WC1E 6BT, UK; [2]Present address: London Pain Consortium, Department of Biological Sciences, South Kensington campus, Imperial College of Science, Technology and Medicine, London, UK; [3]Molecular Nociception Group, Department of Biology, University College London, WC1E 6BT, UK

Introduction

Voltage-gated sodium channels confer excitability on neurons in pain pathways. Because of the recently discovered diversity of sodium channel subtypes, the selective expression of subtypes in nociceptive neurons, and the changes in sodium channel expression that occur in the nervous system after trauma, there is a resurgence of interest in sodium channels as potential drug targets in the treatment of pain. This chapter focuses on sodium channel accessory proteins in pain pathways and their roles in the modification of channel function, expression, and in the interactions of sodium channels with proteins involved in channel tethering to the cytoskeleton and extracellular matrix. In addition, we review the use of the yeast two-hybrid protein interaction trap in the discovery of accessory proteins.

Sodium channel β-subunits

Voltage-gated sodium channels comprise an α-subunit co-associated with at least one accessory β-subunit. The β-subunits modulate the biophysical properties of the channels to the extent that changes in macroscopic current characteristics can be observed in voltage-clamp. The β-subunits also interact with cytoskeletal and extracellular matrix proteins. β-subunits are homologous to the V-set of the immunoglobulin superfamily, including cell adhesion molecules, and comprise a large extracellular domain that incorporates an IgG loop, a single transmembrane domain and a short intracellular domain [1]. The α-subunit incorporates the aqueous pore, voltage-sensing S4 regions (thought to act as activation gates) and an IFM motif between transmembrane domains 3 and 4 (that acts as an inactivation gate that plugs the pore). β-subunits associate non-covalently (e.g., β1) or covalently by an S–S bond (e.g., β2) with the α-subunit, and are involved in extracellular matrix

Sodium Channels, Pain, and Analgesia, edited by Kevin Coward and Mark D. Baker
© 2005 Birkhäuser Verlag Basel/Switzerland

interactions, interactions with the α-subunit and interactions with intracellular cytoskeletal proteins.

The properties of these accessory factors have been investigated in heterologous expression systems following the purification of two proteins associated with α-subunits, β1 and β2. More recently, molecular cloning has identified a protein extensively similar to β1, named β3 [2], and β2-like protein named β4 [3], as well as a splice variant of β1, β1A that appears to have resulted from an intron retention event [4]. Co-expression of β-subunits with sodium channel α-subunits including $Na_V1.2$ and $Na_V1.4$ in heterologous systems have shown that the peak sodium current increases, the voltage-dependence of activation can be steepened, and the voltage-dependence of inactivation shifted to more negative potentials. This has led to the conclusion that β-subunits are crucial for the assembly, expression and for normal functional modulation of the rat brain sodium channel [4–11]. However, an unequivocal involvement in pain mechanisms has not been demonstrated.

Although there is no direct evidence which suggests involvement of β-subunits in pain mechanisms, their association with and ability to regulate α-subunits suggests that they might play some role in regulating the excitability of axons and neurons in pain pathways. Oh et al. [12] reported that β1-subunit mRNA is expressed in large diameter Aβ fibres of dorsal root ganglia (DRG) but that it is almost absent in small diameter unmyelinated C fibre neurons. Some of these authors also reported that β2-subunit mRNA is absent in cultured DRG neurons [13]. However, this result was contradicted by immunohistochemistry using specific antibodies against β1 and β2, where both β1 and β2 subunit proteins were detected in small, medium, and large diameter sensory neurons [14]. The tetrodotoxin-resistant (TTX-r) channels $Na_V1.8$ and $Na_V1.9$ are known to be expressed either exclusively or selectively in nociceptive primary neurons. β-subunits could therefore be co-expressed with TTX-r sodium channels and regulate their function. The relative expression levels of sodium channel α-subunits in the DRG, as well as in the spinal cord, change in rat models of neuropathic pain [15–17]. Levels of β1 and β2 mRNA in the dorsal horn of the spinal cord are also changed from normal and regulated separately in models of neuropathic pain. At 12–15 days after injury, β1 mRNA levels were raised, whereas β2 mRNA levels fell significantly within laminae I–II on the ipsilateral side of the spinal cord [18]. In human cervical sensory ganglia after spinal root avulsion injury, the expression levels of β1 and β2 subunits decreased significantly along with a reduction of $Na_V1.8$ expression [14].

β1 and β3 subunits shift the inactivation curve of $Na_V1.3$ about 10 mV negative, and slow the repriming rate three-fold (here defined as the rate at which the channels can escape inactivation at –80 mV) [19]. As $Na_V1.3$ expression is increased in DRG correlated with the emergence of a rapidly inactivating and rapidly repriming sodium current in a neuropathic pain model [20], the association between β1 or β3 subunits and $Na_V1.3$ may be a key contributor to the severity of neuropathic pain. β3-subunit mRNA is expressed at high levels in small diameter C fibres in rat DRG, and co-expression of β3-subunit with Na$_V$1.8 in *Xenopus* oocytes increased the peak current amplitude when compared with Na$_V$1.8 expressed alone [21]. A significant increase in β3 mRNA expression can be also detected in small diameter sensory neurons of the ipsilateral DRG in the chronic constriction injury model of neuropathic pain. However, recent results from our group show that Na$_V$1.8 is not affected by co-expression of β-subunits including β3 [22] (Fig. 1), encouraging us to look for other interacting proteins. Co-transfection of β-subunits using lipofectamine does not significantly increase the frequency of functional expression or the rate of current inactivation in response to a depolarizing step, which is unphysiologically slow in COS-7 cells. Furthermore, intranuclear injection of α and β3 subunits, i.e., where the α and β subunits are certainly present together, does not give a different result.

β-subunits act as cell adhesion molecules by their ability to interact homophilically through their extracellular immunoglobulin-like repeats. The cytoskeletal protein ankyrin-G is recruited to the cell surface by interacting β-subunits, where the β-subunits bind ankyrin-G with their short cytoplasmic domains. Ankyrin-G is associated with spectrin–actin networks and also interacts with the L1CAM family of cell adhesion molecules that are integral membrane proteins. Along with the L1CAM family members, neurofascin and NrCAM, ankyrin-G is highly concentrated at nodes of Ranvier and at axon initial segments, allowing for the highest density expression of sodium channels in these regions. β-subunits thus link sodium channel α-subunits indirectly both to the cytoskeleton as well as to extracellular matrix proteins such as tenascin-R (secreted by oligodendrocytes) and contactin. The binding of neuronal sodium channels to extracellular matrix molecules may play a role both in functional regulation and in localizing sodium channels in high density at certain areas of the plasma membrane. A ternary complex including sodium channels, neurofascin/NrCAM and ankyrin-G is thus likely to form in myelinated axons. There is evidence that β1, but not β2 subunits, result in increased cell surface Na$^+$ channel expression. McEwen et al. [23] have taken advantage of the fact that β1 subunits enhance sodium channel expression in a heterologous system (CHL 1610), whereas β2 do not. They reasoned that an interaction between the β-subunit and ankyrin-G, plus an interaction with the extracellular matrix protein contactin (the latter not made by β2), is necessary for the sodium channel density modulatory effect. These authors made β1/β2 subunit chimeras (where the external, internal and transmembrane domains could be exchanged) in order to explore this possibility, and they discovered that full length β1 was necessary for enhancement of the sodium current. Na$_V$1.2 interacts with ankyrin-G, and this interaction is enhanced by β1, but when the interaction between β1 and ankyrin-G is prevented by point mutation, then this enhancement is lost. Most recently, McEwen and Isom [24] have shown that an interaction between β1 and neurofascin (Nf186) resulted in an increased channel density, apparently similar to the effect of the β subunit–contactin interaction. Both the intracellular and extracellular interactions of β1 are therefore critically required for substantial modulation of sodium channel density, and probably underlie the interaction

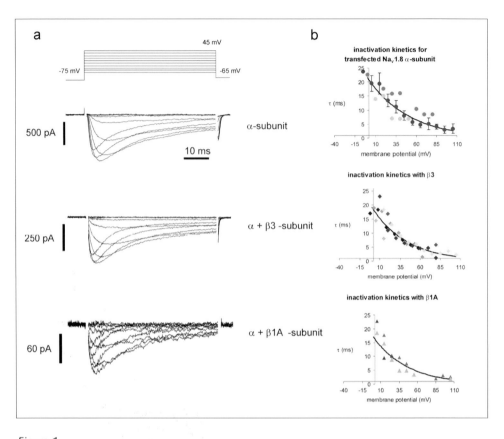

Figure 1
Co-transfection (using lipofectamine) of β-subunits (β1, β1A and β3) does not substantially affect the kinetics of Na_V1.8 sodium currents expressed in COS-7 cells
a) Na_V1.8 sodium currents recorded in voltage-clamp, protocol inset above. In mammalian heterologous systems Na_V1.8 inactivation kinetics are slower than usually found in neurons, and the currents exhibit a more positive activation voltage-dependence. No substantial differences in the biophysical characteristics of the currents are seen with β-subunit co-transfection. b) inactivation time-constant versus membrane potential. Smooth lines are best-fit declining exponentials, e-fold change for α-subunit alone, α + β3 and α + β1A are 50.8, 42.4 and 45.6 mV, respectively. Different cells represented by different grey tones. Means ± s.e.m. plotted for α-subunit alone. The addition of β-subunits does not allow reproduction of neuronal current characteristics, nor does it significantly enhance the frequency of functional transfection (data not shown). We thank Lori Isom for the β1 and β1A-subunit cDNA.

of sodium channels with the extracellular matrix and glial/satellite cells. The same authors also report that β1 and β2 subunits interact extracellularly, where an intracellular sequence of β2 is crucial for this interaction [24].

β-subunit knockouts

The β1-subunit null mutant mouse exhibits a profound phenotype including ataxia and spontaneous seizures [25]. Although there is much evidence that in expression systems β-subunits can alter the voltage-dependence and kinetics of Na$^+$ currents, importantly increasing the rate of inactivation and therefore making the current briefer, one might have expected that inactivation gating would be slowed, and that transient Na$^+$ currents in the brain would be prolonged with the loss of β-subunits. However, knockout of the β1-subunit does not seem to have a widespread effect on sodium current kinetics in the brain, perhaps because other β-subunits can compensate for their loss, and the epileptic phenotype appears to be associated with a change in the levels of expression of Na$_V$1.1 (a decrease) and Na$_V$1.3 (an increase) in discreet areas of the cortex [25]. These findings may help explain the pathology underlying the disease human febrile seizures plus type 1, associated with mutant β1-subunits. Conduction velocities in knock-out optic nerve fibres (including those with the slowest conduction velocities) are reduced, although the most substantial effects are on A-fibres. While pain pathways may conduct more slowly, it is the expression of sodium channels at nodes of Ranvier where an interaction between the channels and contactin is critical, and where the β1 null exhibits a most dramatic functional effect.

The β2-subunit null mutant mouse does not show such a profound phenotype, although β2 is required for normal sodium channel behaviour and expression [26]. Its loss has a more modest effect on the sodium channel expression in brain neuron cell bodies, and at nodes of Ranvier. However, sodium currents recorded in hippocampal neurons are significantly reduced in peak amplitude, and the voltage-dependence of inactivation is shifted more negative.

RPTP-β

It is known that sodium channels are associated with other proteins apart from the β-subunits. For example, receptor protein tyrosine phosphatase-β (RPTP-β) associates with brain neuron sodium channels [27]. RPTP-β has an extracellular (receptor) domain and an intracellular (catalytic) domain, both of which interact with sodium channels. Co-immunoprecipitation experiments revealed that RPTP-β associates with both the α-subunit and β1-subunit, but not with the β2-subunit. In experiments based on the binding properties of β1/β2 subunit chimeras, Ratcliffe et al. found that it is the intracellular region of β1 that binds with RPTP-β [27]. The biophysical properties of Na$_V$1.2 channels are altered by the state of tyrosine phosphorylation, where dephosphorylation increased whole-cell sodium currents by shifting the voltage-dependence of inactivation toward more depolarized potentials. The current amplitude is thus depressed on tyrosine phosphorylation, e.g., by src-kinase, whereas dephosphorylation increases the sodium current [27].

Contactin

Neuropathic pain following nervous system trauma has been associated with the upregulation of $Na_V1.3$, and the downregulation of other sodium channel transcripts, notably $Na_V1.8$ and $Na_V1.9$ ([17] and see Black et al., this volume). For example, $Na_V1.3$ is not normally expressed in the adult rodent DRG, but is expressed following axotomy (although the same may not be true in primates, see Wood, this volume). Furthermore, immunocytochemical evidence indicates that $Na_V1.3$ is expressed at the ends of damaged nerves and in neuromas (e.g., [28]) that are known to be a source of ectopic, spontaneous discharge. Glial cell-line derived neurotrophic factor (GDNF) administration suppresses neuropathic pain behaviour and reverses changes in sodium channel subtype expression [17].

Contactin is a glycosyl-phosphatidylinositol anchored extracellular matrix protein. Shah et al. [28] have reported that co-transfection of $Na_V1.3$ with contactin in human embryonic kidney 293 (HEK293) cells increases the sodium current density three-fold, without affecting the functional properties of the channels. Importantly, the group found that contactin expression was upregulated in axotomized neurons and that the protein accumulated in neuromas. The co-localization of contactin and $Na_V1.3$ in neuromas may thus contribute to the aberrant excitability of damaged nerve, and be a precipitating factor in neuropathic pain, strongly hinting at the involvement of a β-subunit.

Yeast two-hybrid screening against voltage-gated sodium channel α-subunits

Direct interactions of sodium channels with auxiliary β-subunits, RPTP-β and also with tenascin suggests that sodium channels and other molecules involved in action potential generation and propagation might be important players in interactions between proteins that occur during normal neuronal development, the conferment of excitability, and interactions between cells that allow the myelination of axons. One of the techniques that can be used to identify interacting proteins is the yeast two-hybrid interaction trap. The two-hybrid system relies on the fact that eukaryotic transcription factors operate with two separate, and hence modular, domains. One is the DNA-binding domain (DBD) that directs binding to specific DNA sequences and the other is the activating domain that activates transcription [29, 30]. Yeast transcription can be used to assay the interaction between two proteins if one is fused to a DBD and the other fused to an activation domain [31]. Gyuris et al. [32] developed a modification of the two-hybrid system incorporating the following:

A. The bait protein (which is known and in this case is part of a sodium channel), is fused to the DBD. A reporter strain of yeast is transformed with a plasmid that is used to express the part of the sodium channel fused to bacterial transcription factor LexA.

B. A conditionally expressed library cloned in another plasmid is used to transform the yeast strain containing the bait plasmid. A moiety including the nuclear localization signal, transcription activation domain (AD) and epitope tag is fused to the amino terminal of cDNA-encoded proteins. The expression of the resulting hybrid proteins that incorporate the AD is conditional on the presence of galactose and expression is repressed by glucose.

C. The yeast strain used expresses two reporter genes. The LexA binding sites are upstream of these two genes, so that their expression depends on the binding of the hybrid bait protein, and the AD-cDNA fusion protein, that has bound with the bait.

The following proteins that associate with neuronal sodium channels have been identified by others using this approach: calmodulin, syntrophin, fibroblast growth factor homologous factor 1B (FHF1B) and contactin. A new interaction between the sodium channel C-terminal domain and calmodulin (CaM) has been found, by applying the yeast two-hybrid screening method using an expression cDNA rat brain library to the cytoplasmic C-terminal domain of $Na_V1.2$ [33]. The interaction between CaM and other voltage-gated sodium channels were later found to include $Na_V1.4$ [34, 35], $Na_V1.5$ [36], $Na_V1.6$ [35], and $Na_V1.8$ [37]. CaM is an intracellular calcium sensor that binds the ion and subsequently interacts with other molecules, including, e.g., ion channels and calcium/CaM-dependent protein kinase [38]. Although there is no direct evidence that CaM alone is involved in pain pathways, association with CaM is important for functional expression of $Na_V1.4$ and $Na_V1.6$ [35], and CaM may also regulate $Na_V1.8$ expression. However, there is evidence that calcium/CaM-dependent protein kinase II may play a role in pain pathways [39].

The yeast two-hybrid interaction trap and glutathione S-transferase pull-down experiments have indicated that syntrophin γ2 (a scaffolding protein incorporating a PDZ domain), interacts directly with the C terminus of $Na_V1.5$ in [40]. When co-transfected with $Na_V1.5$ into HEK293 cells, syntrophin γ2 affects the voltage-dependence of activation, shifting the activation curve to more positive potentials, and while there appears to be no effect on the steady-state voltage-dependence of inactivation, inactivation kinetics are slowed. Sodium channels in human smooth muscle and cardiac muscle cells exhibit mechanosensitivity, and this is lost when the C terminus-syntrophin γ2 PDZ domain interaction is prevented using competing peptides directed against either region, presumably indicating a loss of the connection between sodium channels and the cytoskeleton.

Liu et al. showed that FHF1B binds with the C-terminal domain of TTX-resistant sodium channel $Na_V1.9$ [41]. This is of potential significance for pain pathways, because $Na_V1.9$ is expressed in nociceptive primary neurons. However, this was not true for other sodium channels known to play important roles in pain pathways. There was no interaction with the C termini of $Na_V1.7$ or $Na_V1.8$, but FHF1B did bind the cardiac sodium channel, $Na_V1.5$, and modulate its functional properties [42]. The voltage-dependence of channel activation and inactivation are both shifted significantly in the hyperpolarizing direction by the binding of FHF1B with $Na_V1.5$ when expressed in HEK293 cells. A mutation in the sodium channel (D1790G) that underlies a long QT interval (LQT-3) phenotype also prevents the interaction of the $Na_V1.5$ channel with FHF1B. This association therefore appears to be vital for normal propagation of the ventricular action potential. Although there is evidence that FHF1B modulates the properties of $Na_V1.5$, the functional significance of the interaction with $Na_V1.9$ remains unresolved.

Some of the same authors [43] reported that the cell adhesion molecule contactin interacted with $Na_V1.9$ at the C-terminal domain of the α-subunit. Contactin is anchored to the membrane through glycerol-phosphatidylinositol, and is entirely extracellular, making a direct interaction with the sodium channel of uncertain physiological importance. However, contactin increased the membrane expression of $Na_V1.9$ in Chinese hamster ovary (CHO) cells when co-transfected with the α-subunit, when compared with transfection of the α-subunit alone. As contactin binds directly to $Na_V1.9$, one possibility is that it may participate in the surface localization of this channel along nociceptive fibres. In comparison, contactin associates with $Na_V1.2$ through the β1 subunit, and increases surface expression by stabilizing the channels in the membrane (e.g., [23]).

Yeast two-hybrid interaction trap and $Na_V1.8$

Using a rat dorsal root ganglion cDNA library, we carried out yeast two-hybrid screening against the five large intracellular domains of $Na_V1.8$. One identified clone encoded annexin II light chain (p11), and this interactor has particularly striking properties. It binds directly to the amino terminus of $Na_V1.8$ and produces functional channels by promoting the translocation of $Na_V1.8$ to the plasma membrane (Fig. 2). Without p11, functional channel expression is very poor [44], and no sodium currents are recorded in CHO-SNS22 cells (a cell line permanently transfected with $Na_V1.8$). When co-expressed with β-subunits $Na_V1.8$ is poorly expressed in cell lines and in *Xenopus* oocytes [26, 45], and this makes the action of p11 all the more remarkable. We found that the endogenous $Na_V1.8$ current in sensory neurons is significantly reduced by injecting vectors incorporating antisense to p11, suggesting that the level of p11 expression in neurons may have important consequences for their firing properties [44]. The binding of $Na_V1.8$ to p11 occurs in a random coiled region flanked by two EF hand motifs whose crystal structure is known. The residues involved are 74–103 of $Na_V1.8$ and 33–78 of p11. Another remarkable finding is that p11 binds to $Na_V1.8$ selectively, and does not bind with other sodium channel subtypes (i.e., $Na_V1.2$, 1.5, 1.7 or $Na_V1.9$) [46].

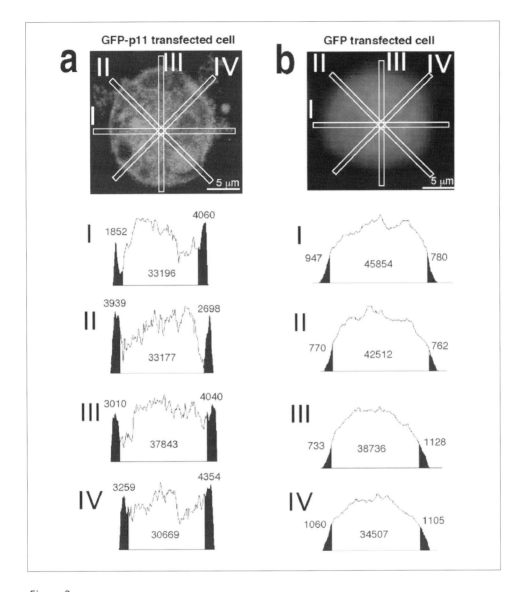

Figure 2
p11 regulates trafficking of Na$_V$1.8 from the cytosol to the membrane of CHO-SNS22 cells, a cell line permanently expressing Na$_V$1.8

a,b, Four typical images of Na$_V$1.8 immunoreactivity obtained from GFP-p11 fusion protein or GFP only in CHO-SNS22 cells, confocal photomicrographs. Density of fluorescence measured along 4 axes, 45° apart through the whole cross section of the cell. Note that in the presence of GFP-p11, the Na$_V$1.8 immunoreactivity is conspicuously concentrated at the membrane relative to controls (with permission, from [44]).

In total, we found 28 different clones that encoded proteins interacting with the intracellular domains of Na$_V$1.8 [37]. Using *in situ* hybridization it became clear that many of these clones exhibiting interactions with Na$_V$1.8 are expressed at high levels in small diameter DRG neurons and the possibility of real, functional interactions were confirmed using immunoprecipitation (pull-down) assays. These include cytoplasmic elements, enzymes, channels, motor proteins, calmodulin and presently unknown proteins (listed in [37]).

Conclusions

Sodium channels are important transmembrane proteins that underlie membrane excitability, including the excitability of neurons in pain pathways. The biophysical properties and densities of sodium channels are modulated by the presence of accessory β-subunits, with the intracellular and extracellular binding properties of the β1-subunit being particularly important in node of Ranvier formation. Other proteins interact with sodium channels, some in a remarkably sub-type selective way. p11 (annexin II light chain) chaperones Na$_V$1.8 to the membrane and plays a crucial role in functional expression. Disrupting p11–Na$_V$1.8 interactions may provide a new way of lowering the expression of TTX-resistant sodium channels in nociceptive neurons, and thus producing analgesia.

Acknowledgements
The authors acknowledge the support of the Wellcome Trust and the MRC.

References

1 Isom LL, Catterall WA (1996) Na$^+$ channel subunits and Ig domains. *Nature* 383: 307–308

2 Morgan K, Stevens EB, Shah B, Cox PJ, Dixon AK, Lee K, Pinnock RD, Hughes J, Richardson PJ, Mizuguchi K et al (2000) β3: an additional auxiliary subunit of the voltage-sensitive sodium channel that modulates channel gating with distinct kinetics. *Proc Natl Acad Sci USA* 97: 2308–2313

3 Yu FH, Westenbroek RE, Silos-Santiago I, McCormick KA, Lawson D, Ge P, Ferriera H, Lilly J, Di Stefano PS, Catterall WA et al (2003) Sodium channel β4, a new disulfide-linked auxiliary subunit with similarity to β2. *J Neurosci* 23: 7577–7585

4 Kazen-Gillespie KA, Ragsdale DS, D'Andrea MR, Mattei LN, Rogers KE, Isom LL (2000) Cloning, localization, and functional expression of sodium channel β1A subunits. *J Biol Chem* 275: 1079–1088

5 Isom LL, De Jongh KS, Patton DE, Reber BF, Offord J, Charbonneau H, Walsh K,

Goldin AL, Catterall WA (1992) Primary structure and functional expression of the β1 subunit of the rat brain sodium channel. *Science* 256: 839–842

6   Isom LL, Ragsdale DS, De Jongh KS, Westenbroek RE, Reber BF, Scheuer T, Catterall WA (1995) Structure and function of the β2 subunit of brain sodium channels, a transmembrane glycoprotein with a CAM motif. *Cell* 83: 433–442

7   Isom LL, Scheuer T, Brownstein AB, Ragsdale DS, Murphy BJ, Catterall WA (1995) Functional co-expression of the β1 and type IIAα-subunits of sodium channels in a mammalian cell line. *J Biol Chem* 270: 3306–3312

8   Patton DE, Isom LL, Catterall WA, Goldin AL (1994) The adult rat brain β1 subunit modifies activation and inactivation gating of multiple sodium channel α-subunits. *J Biol Chem* 269: 17649–17655

9   Schreibmayer W, Wallner M, Lotan I (1994) Mechanism of modulation of single sodium channels from skeletal muscle by the β1-subunit from rat brain. *Pflugers Arch* 426: 360–362

10   Smith MR, Smith RD, Plummer NW, Meisler MH, Goldin AL (1998) Functional analysis of the mouse Scn8a sodium channel. *J Neurosci* 18: 6093–6102

11   Smith RD, Goldin AL (1998) Functional analysis of the rat I sodium channel in *Xenopus* oocytes. *J Neurosci* 18: 811–820

12   Oh Y, Sashihara S, Black JA, Waxman SG (1995) Na⁺ channel β1 subunit mRNA: differential expression in rat spinal sensory neurons. *Mol Brain Res* 30: 357–361

13   Black JA, Dib-Hajj S, McNabola K, Jeste S, Rizzo MA, Kocsis JD, Waxman SG (1996) Spinal sensory neurons express multiple sodium channel α-subunits mRNAs. *Mol Brain Res* 43: 117–131

14   Coward K, Jowett A, Plumpton C, Powell A, Birch R, Tate S, Bountra C, Anand P (2001) Sodium channel β1 and β2 subunits parallel SNS/PN3 α-subunit changes in injured human sensory neurons. *Neuroreport* 12: 483–488

15   Waxman SG, Cummins TR, Dib-Hajj S, Fjell J, Black JA (1999) Sodium channels, excitability of primary sensory neurons, and the molecular basis of pain. *Muscle Nerve* 22: 1177–1187

16   Okuse K, Chaplan SR, McMahon SB, Luo ZD, Calcutt NA, Scott BP, Akopian AN, Wood JN (1997) Regulation of expression of the sensory neuron-specific sodium channel SNS in inflammatory and neuropathic pain. *Mol Cell Neurosci* 10: 196–207

17   Boucher TJ, Okuse K, Bennett DL, Munson JB, Wood JN, McMahon SB (2000) Potent analgesic effects of GDNF in neuropathic pain states. *Science* 290: 124–127

18   Blackburn-Munro G, Fleetwood-Walker SM (1999) The sodium channel auxiliary subunits β1 and β2 are differentially expressed in the spinal cord of neuropathic rats. *Neuroscience* 90: 153–164

19   Meadows LS, Chen YH, Powell AJ, Clare JJ, Ragsdale DS (2002) Functional modulation of human brain Na$_V$1.3 sodium channels, expressed in mammalian cells, by auxiliary β1, β2 and β3 subunits. *Neuroscience* 114: 745–753

20   Black JA, Cummins TR, Plumpton C, Chen YH, Hormuzdiar W, Clare JJ, Waxman SG

(1999) Upregulation of a silent sodium channel after peripheral, but not central, nerve injury in DRG neurons. *J Neurophysiol* 82: 2776–2785

21 Shah BS, Stevens EB, Gonzalez MI, Bramwell S, Pinnock RD, Lee K, Dixon AK (2000) β3, a novel auxiliary subunit for the voltage-gated sodium channel, is expressed preferentially in sensory neurons and is upregulated in the chronic constriction injury model of neuropathic pain. *Eur J Neurosci* 12: 3985–3990

22 Baker MD, Poon W-YL, Wood JN, Okuse K (2004) Functional effects of co-transfecting β-subunits 1, 1A and 3 with Na$_V$1.8 α-subunit in a COS-7 heterologous system. *J Physiol* 555P: PC20

23 McEwen DP, Meadows LS, Chen C, Thyagarajian V, Isom L (2004) Sodium channel β1 subunit-mediated modulation of Na$_V$1.2 currents and cell surface density is dependent on interactions with contactin and ankyrin. *J Biol Chem* 279: 16044–16049

24 McEwen DP, Isom L (2004) Heterophilic interactions of sodium channel β1 subunits with axonal and glial cell adhesion molecules. *J Biol Chem* 279: 52744–52752

25 Chen C, Westenbroek RE, Xu X, Edwards CA, Sorenson DR, Chen Y, McEwen DP, O'Malley HA, Bharucha V, Meadows LS et al (2004) Mice lacking sodium channel β1 subunits display defects in neuronal excitability, sodium channel expression, and nodal architecture. *J Neurosci* 24: 4030–4042

26 Chen C, Bharucha V, Chen Y, Westenbroek RE, Brown A, Malhotra JD, Jones D, Avery C, Gillespie PJ 3rd, Kazen-Gillespie KA et al (2002) Reduced sodium channel density, altered voltage dependence of inactivation, and increased susceptibility to seizures in mice lacking sodium channel β2-subunits. *PNAS* 99: 17072–17077

27 Ratcliffe CF, Qu Y, McCormick KA, Tibbs VC, Dixon JE, Scheuer T, Catterall WA (2000) A sodium channel signaling complex: modulation by associated receptor protein tyrosine phosphatase β. *Nat Neurosci* 3: 437–444

28 Shah BH, Rush AM, Liu S, Tyrell L, Black JA, Dib-Hajj SD, Waxman SG (2004) Contactin associates with sodium channel Na$_V$1.3 in native tissues and increases channel density at the cell surface. *J Neurosci* 24: 7387–7399

29 Hope IA, Struhl K (1986) Functional dissection of a eukaryotic transcriptional activator protein GCN4 of yeast. *Cell* 46: 885–894

30 Keegan L, Gill G, Ptashne M (1986) Separation of DNA binding from the transcription-activating function of a eukaryotic regulatory protein. *Science* 231: 699–704

31 Field S, Song O (1989) A novel genetic system to detect protein-protein interactions. *Nature* 340: 245–246

32 Gyuris J, Golemis E, Chertkov H, Brent R (1993) Cdi1, a human G1 and S phase protein phosphatase that associates with cdk2. *Cell* 75: 791–803

33 Mori M, Konno T, Ozawa T, Murata M, Imoto K, Nagayama K (2000) Novel interaction of the voltage-dependent sodium channel (VDSC) with calmodulin: does VDSC acquire calmodulin-mediated sensitivity? *Biochemistry* 39: 1316–1323

34 Deschenes I, Neyroud N, DiSilvestre D, Marban E, Yue DT, Tomaselli GF (2002) Isoform-specific modulation of voltage-gated Na$^+$ channels by calmodulin. *Circ Res* 90: E49–57

35 Herzog RI, Liu L, Waxman SG, Cummins TR (2003) Calmodulin binds to the C terminus of sodium channels Na$_V$1.4 and Na$_V$1.6 and differentially modulates their functional properties. *J Neurosci* 23: 8261–8270

36 Tan HL, Kupershmidt S, Zhang R, Stepanovic S, Roden DM, Wilde AA, Anderson ME, Balser JR (2002) A calcium sensor in the sodium channel modulates cardiac excitability. *Nature* 415: 442–447

37 Malik-Hall M, Poon W-YL, Baker MD, Wood JN, Okuse K (2002) Sensory neuron proteins interact with the intracellular domains of sodium channel Na$_V$1.8. *Mol Brain Res* 110: 298–304

38 Nelson MR, Chazin WJ (1998) In: Van Eldik LJ, Watterson DM (eds): *Calmodulin and signal transduction.* Academic Press, San Diego, CA, 17–64

39 Fang L, Wu J, Lin Q, Willis WD (2002) Calcium-calmodulin-dependent protein kinase II contributes to spinal cord central sensitization. *J Neurosci* 22: 4196–4204

40 Ou Y, Strege P, Miller SM, Makielski J, Ackerman M, Gibbons SJ, Farrugia G (2003) Syntrophin γ2 regulates SCN5A gating by a PDZ domain-mediated interaction. *J Biol Chem* 278: 1915–1923

41 Liu C, Dib-Hajj SD, Waxman SG (2001) Fibroblast growth factor homologous factor 1B binds to the C terminus of the tetrodotoxin-resistant sodium channel rNa$_V$1.9a (NaN). *J Biol Chem* 276: 18925–18933

42 Liu CJ, Dib-Hajj SD, Renganathan M, Cummins TR, Waxman SG (2002) Modulation of the cardiac sodium channel Na$_V$1.5 by fibroblast growth factor homologous factor 1B. *J Biol Chem* 278: 1029–1036

43 Liu C, Dib-Hajj SD, Black JA, Greenwood J, Lian Z, Waxman SG (2001) Direct interaction with contactin targets voltage-gated sodium channel Na$_V$1.9/NaN to the cell membrane. *J Biol Chem* 276: 46553–46561

44 Okuse K, Malik-Hall M, Baker MD, Poon W-YL, Kong H, Chao MV, Wood JN (2002) The annexin II light chain p11 regulates the functional expression of the sensory neuron specific sodium channel. *Nature* 417: 653–656

45 Sangameswaran L, Delgado SG, Fish LM, Koch BD, Jakeman LB, Stewart GR, Sze P, Hunter JC, Eglen RM, Herman RC (1996) Structure and function of a novel voltage-gated, tetrodotoxin-resistant sodium channel specific to sensory neurons. *J Biol Chem* 271: 5953–5956

46 Poon WY, Malik-Hall M, Wood JN, Okuse K (2004) Identification of binding domains in the sodium channel Na$_V$1.8 intracellular N-terminal region and annexin II light chain p11. *FEBS Lett* 558: 114–118

Sodium channels and nociceptive nerve endings

James A. Brock

Prince of Wales Medical Research Institute, Barker St, Randwick, Sydney NSW 2031, Australia

Introduction

The mechanisms whereby sensory stimuli applied to receptive nerve endings of nociceptive neurones are transformed into action potentials that propagate centrally to elicit painful sensations remain largely a matter of conjecture. It is assumed that the receptive region of the axon at the nerve terminal is functionally and spatially separated from the site at which action potentials are initiated. The receptive region of the axon contains the primary detector molecules (ion channels and/or G-protein coupled receptors) and is normally considered to be devoid of Na^+ channels. The depolarization (receptor potential) generated by the sensory stimulus in the receptive region of the axon propagates passively (electrotonically) along the axon to a more proximal point where action potentials are initiated. The axonal membrane at this site is predicted to have a relatively high density of Na^+ channels, which gives this region of the axon the lowest voltage threshold for action potential initiation.

The primary purpose of this chapter is to consider the properties of the Na^+ channels expressed in unmyelinated (C) and thinly myelinated ($A\delta$) nociceptive neurones that are likely to contribute to regulating the excitability of the sensory nerve endings under normal conditions and following interventions that result in increased excitability.

In dorsal root ganglia (DRG), the cell bodies of C neurones have small diameters ($< 25 \ \mu m$) and those of $A\delta$ neurones have small to medium diameters ($< 45 \ \mu m$). In rats, the largest sub-population of cutaneous C neurones consists of the polymodal nociceptors that respond to noxious mechanical, thermal and chemical stimuli. The population of cutaneous C neurones also contains other subgroups of nociceptors including high threshold mechano-receptors and cold nociceptors. Cutaneous $A\delta$ neurones include high threshold mechano-heat receptors as well as neurones responding to non-noxious mechanical or thermal stimuli (warm and cold receptors). It is likely that most $A\delta$ high threshold mechano-heat receptors can also respond to noxious chemical stimuli [1].

Sodium Channels, Pain, and Analgesia, edited by Kevin Coward and Mark D. Baker
© 2005 Birkhäuser Verlag Basel/Switzerland

Voltage-gated Na$^+$ channels expressed in small and medium sized sensory neurones

Voltage-gated Na$^+$ channels (VGSC) are made up of an α-subunit and one or more β-subunits. The α-subunit alone forms functional Na$^+$ selective channels and contains the channel pore, the voltage sensor and the structural elements responsible for fast inactivation. The β-subunits are suggested to serve a number of functions, including modulation of α-subunit function and targeting/anchoring the channels at specific sites in the plasma membrane. To date, nine genes encoding α-subunits (Na$_V$1.1–Na$_V$1.9) and three genes encoding β-subunits (β1–3) have been identified.

Studies using *in situ* hybridization in adult rats have revealed that almost all DRG neurones express detectable levels of mRNA for the α-subunits Na$_V$1.6 and Na$_V$1.7, although the level of Na$_V$1.6 expression is lower in small diameter neurones [2]. Messenger RNA for Na$_V$1.8 is highly expressed in many of the small diameter neurones and is also expressed in a subpopulation of the medium diameter neurones [2–4]. Many of the small diameter neurones also express mRNA for Na$_V$1.9 [5]. Hybridization signal for all three β-subunits has been identified in DRG neurones, although relatively few neurones contain detectable levels of mRNA for the β2 subunit [6, 7]. Both small and medium diameter neurones express mRNA for the β3 subunit whereas mRNA for the β1 subunit is expressed in many of the medium sized neurones and is rarely detected in small diameter neurones.

The immunohistochemical detection of Na$_V$1.8 and Na$_V$1.9 proteins in adult rat DRG neurones correlates closely with the expression of their mRNA [8, 9]. About 50% of C neurones express detectable levels of Na$_V$1.8 and Na$_V$1.9 protein, and in most of these neurones the subunits are co-localized. Many of the C neurones expressing Na$_V$1.8 and/or Na$_V$1.9 protein (~60%) also express the vanilloid receptor TRP$_V$1 (or VR1), indicating that they are likely to be heat-sensitive or polymodal nociceptors. Only about 10% of Aδ neurones express Na$_V$1.8 protein and these all express the nerve growth factor (NGF) receptor Trk-A and many express the vanilloid-like receptor VRL1 [8], which is thought to be responsible for transducing high-threshold heat responses in Aδ nociceptive neurones [10]. While mRNA for Na$_V$1.7 is uniformly expressed in the majority of DRG neurones, anti-Na$_V$1.7 antibodies show more intense labeling of small than large sized DRG neurones [11]. Similarly, Na$_V$1.6 protein is present in all DRG neurones, with antibody labeling most pronounced in small diameter neurones [12].

Correlation of sensory receptor properties of rat DRG neurones with expression of α-subunits has revealed that about 90% of C and Aδ neurones identified as nociceptors were immunoreactive for Na$_V$1.8 [13]. Weak labeling for Na$_V$1.8 was also detected in C and Aδ neurones identified as low threshold mechano-receptive units. Na$_V$1.9 was exclusively expressed in neurones classified as nociceptors, with immunoreactivity being detected in about 60% of C and Aδ nociceptive units [14]. In guinea pig DRG, Na$_V$1.7 immunoreactivity was detected in all C neurones and

90% of Aδ neurones tested, including both nociceptive and low threshold mechano-receptive units [15].

Knowledge of the distribution of Na$^+$ channel α-subunits in the peripheral and centrally projecting axons of rat sensory neurones remains rudimentary. Na$_V$1.6 is the predominant Na$^+$ channel expressed at nodes of Ranvier in myelinated sensory axons [16]. The Na$_V$1.6 subunit is also present throughout the peripheral axon of unmyelinated sensory neurones [17]. Within the rat cornea, Na$_V$1.8 and Na$_V$1.9 are expressed along the entire length of the unmyelinated sensory axons including the nerve terminals in the most superficial layer of the corneal epithelium [18, 19]. Within the sciatic nerve, Na$_V$1.9 is preferentially located in isolectin B4 positive unmyelinated axons and at the nodes of Ranvier of some thinly myelinated axons [19]. In rat DRG, isolectin B4 binds to a subset of nociceptive C neurones [20]. The distribution of Na$_V$1.7 along the axons of sensory neurones has not been reported.

Importantly, the distribution of antibody binding for Na$_V$1.7, Na$_V$1.8 and Na$_V$1.9 in human DRG neurones closely parallels that observed in rat DRG neurones [21, 22]. Labeling for Na$_V$1.7 was observed in 80–90% of small, medium and large diameter neurones, but the most intense labeling was obtained in the small diameter neurones. All small diameter neurones were positive for Na$_V$1.8 and Na$_V$1.9, but these subunits were also detected in 60–80% of medium and large diameter neurones. Labeling for Na$_V$1.8 and Na$_V$1.9 was most intense in the small diameter neurones.

Electrophysiological investigation of Na$^+$ channels

In general, Na$^+$ channel α-subunits can be divided into two groups based on their sensitivity to tetrodotoxin (TTX). For those expressed at detectable levels in many of the small and/or medium sized neurones in adult rat DRG, Na$_V$1.6 and Na$_V$1.7 are readily blocked by TTX (IC$_{50}$ < 10 nM) [23–26] whereas Na$_V$1.8 is resistant to blocking actions of this agent (IC$_{50}$ > 10 μM) [3, 4, 27]. On the basis of its amino acid sequence, Na$_V$1.9 is predicted to encode a TTX-resistant Na$^+$ channel [5, 27]. A range of local anaesthetics and antiepileptic agents block both TTX-sensitive and TTX-resistant Na$^+$ channels but none have sufficient selectivity to discriminate between the different types of Na$^+$ channel.

Studies of DRG neurones have identified cells with only a TTX-sensitive Na$^+$ current or a TTX-resistant current as well as cells that display both types of current. Typically the TTX-sensitive current has a low voltage threshold for activation (~–50 mV) and has fast activation and inactivation [28–30]. Once inactivated, the Na$^+$ channels producing the TTX-sensitive current recover from inactivation relatively slowly. As the mid-point of steady state inactivation for the TTX-sensitive Na$^+$ current is near –70 mV, at the normal resting membrane potentials (~–60 mV) less than 50% of the channels underlying the TTX-sensitive current are available for activation.

At least two distinct types of TTX-resistant Na⁺ current have been described in adult sensory neurones. The TTX-resistant Na⁺ current most commonly reported in small and medium sized sensory neurones has a relatively high voltage threshold for activation (~–30 mV) and activates and inactivates relatively slowly [29, 30]. However, the ion channels producing this Na⁺ current recover from inactivation very rapidly. As the mid-point of steady state inactivation for this TTX-resistant current is in the range –40 to –25 mV, at the normal resting membrane potentials nearly 100% of these channels would be available for activation. This Na⁺ current, which is believed to generate the TTX-resistant action potentials recorded in adult sensory neurones (see below), has properties similar to those produced by $Na_V1.8$ α-subunits expressed in *Xenopus* oocytes [3, 4] and is absent in sensory neurones from $Na_V1.8$-null mice [31]. Importantly, nuclear injection of $Na_V1.8$ cDNA into sensory neurones from $Na_V1.8$-null mice restores this TTX-resistant Na⁺ current [31].

The second TTX-resistant Na⁺ current is recorded predominantly in small diameter sensory neurones. It has a low voltage threshold for activation (~–70 mV) and activates and inactivates very slowly with a mid-point for steady state inactivation at ~–45 mV [32]. As there is a substantial overlap between the activation and steady state inactivation curves, the channels responsible for this TTX-resistant Na⁺ current would be expected to produce a persistent window current at the normal resting membrane potential. The very slow activation of these Na⁺ channels makes it unlikely that they contribute to generation of action potentials. However, the persistent current produced by these Na⁺ channels has been proposed to play an important role in controlling the resting membrane potential [33]. As this TTX-resistant Na⁺ current is present in small diameter neurones from wild type and $Na_V1.8$-null mice [32], $Na_V1.9$ is suggested to be responsible for this persistent current.

Functional roles of TTX-sensitive and TTX-resistant resistant Na⁺ channels

Using current injection into the cell body, the sensory neurones in intact rodent DRG and trigeminal ganglia can be divided into those with TTX-sensitive action potentials and those with TTX-resistant action potentials [34, 35]. In general, neurones with medium to large diameter cell bodies generate TTX-sensitive action potentials whereas neurones with small diameters generate TTX-resistant action potentials (Fig. 1Aa and Ca). However, there is also a subset of Aδ neurones with medium sized cell bodies that are able to support TTX-resistant action potentials (Fig. 1Ba) [36]. Typically, TTX-sensitive action potentials are fast (duration ≤ 1 ms) whereas TTX-resistant action potentials are slow (duration ≥ 2 ms) and a have characteristic hump (inflexion) on their falling phase (Fig. 1B and C).

In contrast to the action potentials evoked by direct current injection into the cell body of rat DRG neurones, those elicited by antidromic invasion of the cell body following excitation of the peripheral axon are blocked by TTX in all sensory neu-

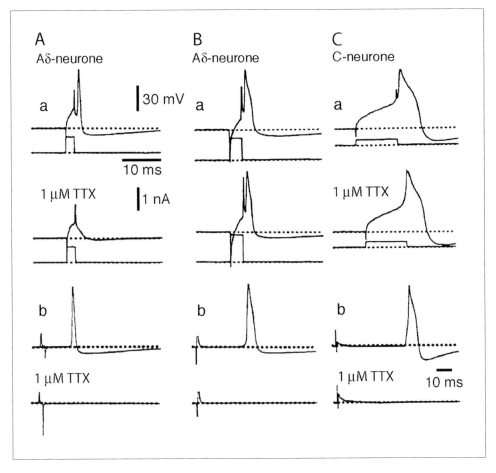

Figure 1
Sensitivity of somatic and conducted action potentials recorded in Aδ (A and B) and C (C) dorsal root ganglion (L4/L5) neurones to tetrodotoxin (TTX, 1 μM)
Aa, Ba and Ca: action potentials (upper traces) evoked by current injection (lower traces) into the cell body in control solution (top) and in TTX (bottom). Ab, Bb and Cb: responses evoked by electrical excitation of sciatic axons in control solution (top) and in TTX (bottom). In all neurones the action potentials evoked by stimulating the sciatic nerve were blocked by TTX (Ab, Bb and Cb). This finding indicates that conduction along the peripheral axon is dependent on TTX-sensitive Na⁺ channels. In response to current injection into the cell body, Aδ neurones had either fast TTX-sensitive action potentials (Aa) or slow TTX-resistant action potentials (Ba) whereas all C neurones tested had slow TTX-resistant action potentials (Ca). In both cell types, the TTX-resistant action potentials have a characteristic hump (inflexion) on their falling phase. The calibration in Aa applies to all records except for the time calibration in Cb. Used by permission [36]

rones (Fig. 1Ab, Bb and Cb) [34, 36, 37]. Therefore the density of TTX-resistant Na⁺ channels along the peripheral axons is normally insufficient to support action potentials. In biopsies of human sural nerve from patients with a range of neurological conditions, there is a substantial component of the C-fibre compound action potential that is due to activation of TTX-resistant Na⁺ channels [38]. However, in this study, it is possible that TTX-resistance of action potentials was produced by nerve pathology [22, 39].

Evidence for TTX-resistant action potentials in sensory nerve terminals

The mechanisms controlling the excitability of nociceptor terminals are largely a matter of speculation because of their size (< 0.5 μm diameter) and inaccessibility in intact tissues like skin. What is known has been inferred indirectly from recordings of discharge of afferent axons when the environment of the receptor is pharmacologically manipulated. To investigate directly the role of TTX-sensitive and TTX-resistant Na⁺ channels in regulating the excitability of nociceptive nerve endings, we recently developed an extracellular recording technique that allows electrical activity to be recorded from single sensory nerve terminals in the guinea pig cornea [40].

The cornea is very densely supplied by small diameter sensory nerve endings that terminate abruptly as they approach the most superficial layer of the corneal epithelium. Recordings from the ciliary nerves at the back of the eye have identified three types of sensory receptors (polymodal, mechano-sensory and cold-sensitive) in the cornea [41]. Using a small diameter (≤ 50 μm) suction electrode applied to the epithelial surface of the cornea, nerve impulses originating in single sensory nerve terminals can be recorded (Fig. 2). Using locally applied stimuli, these nerve terminals can be identified as either polymodal receptors or cold-sensitive receptors. The conduction velocities of these sensory axons range from 0.3–2.7 m s⁻¹ (Fig. 2D). Mechano-sensory units identified in the ciliary nerves at the back of the eye have conduction velocities approaching 10 m s⁻¹. Such mechano-sensory nerve terminals have not been identified at the surface of the corneal epithelium.

In the cornea, polymodal receptors are nociceptive whereas cold receptors, when stimulated by small decreases in temperature, give rise to the sensation of cooling [42]. However, during more pronounced reductions in temperature, co-activation of polymodal and cold receptors may contribute to a sensation of irritation [43].

For all sensory nerve terminals identified using the extracellular recording technique, bath application of TTX (1 μM) blocked nerve impulses evoked by electrical stimulation of the ciliary nerves at the back of the eye (Fig. 3) [40]. This finding indicates that action potential conduction along the main axon is dependent on activation of TTX-sensitive Na⁺ channels. However, in both polymodal and cold-sensitive receptors, ongoing and/or sensory stimulus-evoked nerve impulses persisted in the presence of TTX (Figs. 3 and 4). Bath application of the local anaesthetic, lignocaine

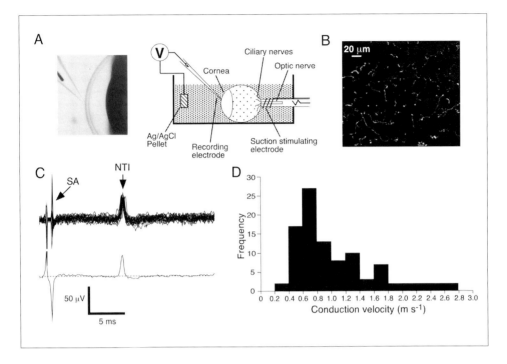

Figure 2
Recording from the corneal epithelium
A: schematic diagram of recording set-up and photomicrograph showing the location of the recording electrode (scale bar, 1 mm). B: confocal micrograph of nerve terminals revealed with antibody to PGP 9.5 in the guinea-pig cornea. Most nerve terminals approach the surface of the epithelium at right angles and appear as single dots. C: a single nerve terminal impulse (NTI) evoked by electrical stimulation of the ciliary nerves at the back of the eye. The upper part shows 50 overlaid traces recorded during a train of stimuli at 1 Hz and the lower part shows the average of these traces (SA, stimulation artefact). D: frequency distribution of conduction velocities for all single units recorded. Used by permission [40]

(1 mM), which is known to block both the TTX-sensitive and TTX-resistant current in cell bodies of sensory neurones [44], blocked all electrical activity. These findings indicate that TTX-resistant Na⁺ channels play a major role in initiating action potentials in the sensory nerve terminals. A similar conclusion has been made for mechano-sensory C-fibres and slowly conducting Aδ fibres in the dura of the rat [45]. In the latter study, electrical activity was recorded from neurones in the trigeminal ganglion while TTX was applied in the vicinity of the receptive nerve endings in the dura. Mechanically-evoked responses of some C-fibres in rat hind paw skin have also been reported to be resistant to TTX [46].

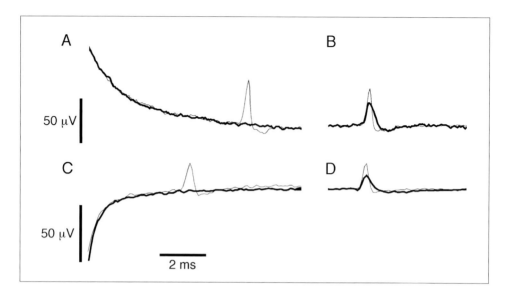

Figure 3
Effects of tetrodotoxin (TTX, 1 µM for 30 min) on electrically-evoked and spontaneously occurring nerve terminal impulses (NTIs) recorded from a polymodal receptor (A and B) and a cold receptor (C and D)
A–D: averaged electrically-evoked (A and C) and spontaneously occurring (B and D) NTIs recorded before (thin line) and in the presence of TTX (thick line). TTX blocked the NTIs evoked by electrical stimulation of the ciliary nerves but did not stop the ongoing nerve activity, indicating that TTX-resistant Na⁺ channels alone can support action potentials in the nerve terminals. TTX did slow the time course of the spontaneously occurring NTIs, indicating that TTX-sensitive Na⁺ channels normally contribute to action potentials in the nerve terminals. Used by permission [40]

Contributions of TTX-sensitive and TTX-resistant Na⁺ channels to nerve terminal action potentials

While bath application of TTX did not stop the ongoing and sensory stimulus-evoked activity in the nerve terminals, it did slow the time course of all nerve impulses indicating that TTX-sensitive Na⁺ channels contribute to action potential generation in the nerve terminals of both polymodal and cold receptors (Fig. 3) [47]. The effects of focally applying lignocaine through the recording microelectrode have also been investigated [47, 48]. In polymodal receptors, focal application of lignocaine slowed the time course of nerve impulses, indicating that the nerve terminals possess sufficient Na⁺ channels to support active propagation of impulses into the

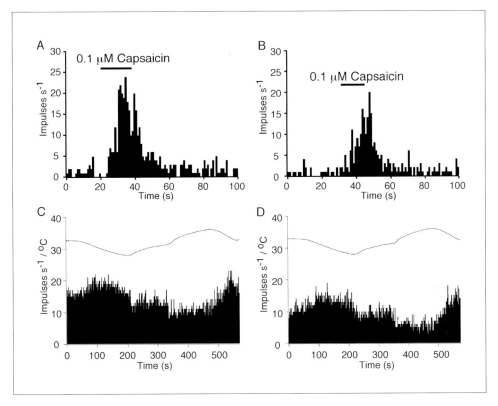

Figure 4
Effects of tetrodotoxin (TTX, 1 μM for 30 min) on sensory stimulus-evoked nerve activity
A and B: the effect of capsaicin (0.1 μM) on the frequency of occurrence of nerve terminal
impulses (NTIs) recorded from a polymodal receptor before (A) and during (B) application of
TTX. C and D: the effect of temperature changes (upper curve) on the frequency of occur-
rence of NTIs recorded from a cold-sensitive receptor before (C) and during (D) application
of TTX. The histograms have bin widths of 1 s. Used by permission [40]

nerve endings (Fig. 5A). This property may be important for the efferent function of
polymodal receptors, active propagation of nerve impulses to other branches within the nerve terminal arbor triggering the release of neuropeptides contained in these
nerve endings. In contrast, local application of lignocaine to cold-sensitive receptors
did not change the time course of nerve impulses (Fig. 5B). This finding suggests that
there is not sufficient Na$^+$ channels available in nerve terminal membrane to support
action potentials and that cold sensitive receptors are passively invaded from a point
more proximal in the axon where action potentials can fail or be initiated.

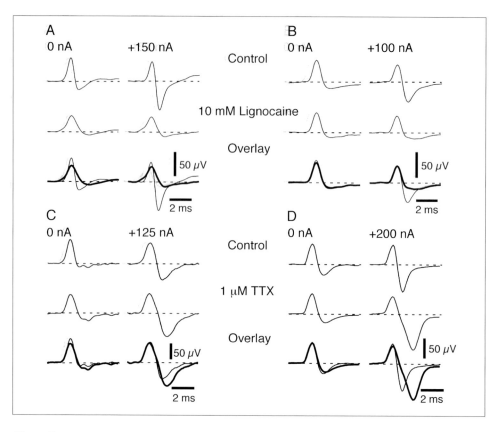

Figure 5
Effects of locally applied lignocaine (10 mM, A and B) and tetrodotoxin (TTX, 1 μM, C and D) on the polarization-induced changes in nerve terminal impulse (NTI) shape in polymodal (A and C) and cold-sensitive receptors (B and D)

Each panel shows averaged NTIs recorded before and during application of lignocaine or TTX. In each case, NTIs were recorded in the absence of polarizing current and during application of hyperpolarizing (+ve) current. The lower records in panels A to C show control and drug treated (thick line) traces overlaid. In the absence of polarizing current, lignocaine produced a pronounced slowing in the time course of NTIs in polymodal receptors but did not change the time course of NTIs in cold receptors. In both types of receptor, lignocaine abolished the increase in the negative component of NTIs produced by hyperpolarizing current. In the absence of polarizing current, TTX had little effect on the time course of NTIs in both the polymodal receptor and the cold receptor and did not inhibit the increase in the negative component of the NTI produced by hyperpolarizing current. In both types of receptor, TTX slowed the time course of the negative component of the NTI revealed by hyperpolarizing current, indicating that TTX-sensitive Na+ channels normally contribute to shaping this component of the signal. Used by permission [48]

The nerve impulses recorded with the extracellular recording technique are proportional to the net membrane current, which is comprised of both capacitive and ionic components. For the recording configuration used in these experiments, positive deflections represent net outward current and negative deflections represent net inward current. Under control conditions, where the biphasic nerve impulses are largely positive going (Figs. 2, 3 and 5), the net membrane current is predominantly outward and is due primarily to the capacitive current generated when the impulse invades the nerve terminal [49]. Consistent with this interpretation, the nerve impulses remain largely positive-going during local application of Na^+ or K^+ channel blockers [48].

Hyperpolarizing the nerve terminal by passing positive current through the recording electrode selectively increases the negative-going component of the nerve impulse in both polymodal and cold receptors (Fig. 5A–D) [48]. This net increase in inward current reflects an increase in Na^+ current as it is markedly reduced when the recording electrode is perfused with low Na^+ solution but is little changed when the solution contained no added Ca^{2+} [48]. The effects of hyperpolarizing the nerve terminal are completely blocked by locally applied lignocaine (Fig. 5A and B). In contrast, the hyperpolarization-induced increase in inward current is resistant to TTX (Fig. 5C and D). In presence of TTX, the nerve impulses recorded during application of hyperpolarizing current often had an inflection close to the point where the signals reversed polarity (Fig. 5D). This change may reflect blockade of TTX-sensitive Na^+ channels that contribute to an early component of the inward Na^+ current (see below). TTX also prolonged the duration of the inward current. These findings indicate that TTX-resistant Na^+ channels make a dominant contribution to the nerve terminal Na^+ current.

In cold receptors, the increase in inward current produced by hyperpolarizing current suggests that the sensory nerve terminals do possess Na^+ channels. However, under normal conditions, these channels do not appear to contribute to the signals recorded. For this reason, it is assumed that the nerve terminals of cold receptors are relatively depolarized and, as a result, most of the Na^+ channels present in the nerve terminal membrane are inactivated.

Functional roles of TTX-sensitive and TTX-resistant Na^+ channels in nerve terminals

Consistent with the findings for polymodal receptors in the cornea, the cell bodies of small diameter sensory neurones express both TTX-sensitive and TTX-resistant Na^+ channels. They therefore provide a reasonable model system in which to investigate the roles of TTX-sensitive and TTX-resistant Na^+ channels in action potential generation in the receptive nerve ending of nociceptors.

As indicated above, at normal resting membrane potentials (~–60 mV) less than 50% of the TTX-sensitive Na+ channels are available for activation whereas most of the TTX-resistant Na+ channels that contribute to the action potential are available for activation. During the upstroke of the action potential, the TTX-sensitive current activates more quickly and at smaller depolarizations than the TTX-resistant current, so that the threshold potential for action potential initiation moves to more depolarized values when this current is blocked [50, 51]. Because the midpoint of steady state inactivation for the TTX-sensitive current is near –70 mV, the contribution of TTX-sensitive Na+ currents to action potential initiation will be very sensitive to changes in the resting membrane potential. If the resting membrane potential is low (~–50 mV), the TTX-sensitive current makes little contribution to the action potential [52].

The TTX-resistant Na+ channels activate rapidly enough to carry the largest inward movement of charge during the upstroke of the action potential [51, 53]. Furthermore, because these channels inactivate relatively slowly, they contribute about 70% of the inward current during the action potential. This finding accords with our finding that TTX-resistant Na+ channels make a dominant contribution to the nerve terminal Na+ current.

The characteristic hump on falling phase of the action potential in small diameter neurones reflects, in part, slow inactivation of the TTX-resistant Na+ current [51]. However, high voltage-activated Ca^{2+} channels also contribute a substantial component of the inward movement of charge during this phase of the action potential [51].

Several features of the Na+ current producing the TTX-resistant action potentials are likely to play a role in determining the behaviour of nociceptive nerve endings. The relatively large depolarization required to activate the TTX-resistant current is likely to contribute to these receptors having high sensory thresholds. Furthermore, as the mid-point of the steady state inactivation curve for the TTX-resistant Na+ current is in the range –40 to –25 mV, these nerve terminals should be able to retain their ability to fire action potentials when the membrane is depolarized by as much as 10–20 mV. The very rapid recovery of the TTX-resistant Na+ current from inactivation ought to enable the nerve terminals to fire long trains of action potentials in response to a maintained depolarizing stimulus. However, the TTX-resistant Na+ current displays a very slow inactivation process (over tens of seconds) that during maintained depolarization may switch the current off [30].

As $Na_V1.9$ is present in peripheral terminals of unmyelinated sensory axons [18, 19], it is likely that the TTX-resistant persistent Na+ current also contributes to the behaviour of nociceptive nerve endings. These channels would be expected to provide a depolarizing current that contributes to setting the resting membrane potential [33]. In addition, increased activity of these channels during slow depolarizations would be expected to amplify these signals, increasing the likelihood that they initiate action potentials [33].

Nociceptor sensitization

Hyperalgesia is a characteristic consequence of inflammation resulting from tissue injury. In the inflamed tissue, this pain reflects an increased excitability of the nociceptive nerve terminals resulting in reduced sensory thresholds. The increase in excitability is produced by the action of a variety of inflammatory mediators that include prostaglandins, bradykinin, 5-hydroxytryptamine, histamine, adenosine, cytokines (e.g., TNFα) and NGF. The mechanisms by which these agents produce this change in excitability are complex and involve multiple signal transduction pathways. In particular, modulation of Na$^+$ channel activity and expression has been implicated as playing a key role in nociceptor sensitization in inflamed tissues.

Modulation of Na$^+$ channel activity

The proinflammatory prostaglandin, prostaglandin E$_2$ (PGE$_2$) has been demonstrated to increase the sensitivity of sensory neurones both *in vivo* [54, 55] and *in vitro* [56, 57]. In behavioural tests, PGE$_2$ lowers the sensory thresholds for both thermal and mechanical stimuli [58, 59]. Sensory neurones are known to express multiple PGE$_2$ receptor subtypes (EP receptors) [60] and the sensitizing action of PGE$_2$ is suggested to occur through a direct action on the sensory neurones. The prime evidence supporting this assertion comes from investigating the effects of PGE$_2$ on small diameter sensory neurones isolated in culture. In neurones isolated from embryonic rat DRG, acute administration of PGE$_2$ increases the number of action potentials elicited by a depolarizing stimulus without changing resting membrane potential or the size of the depolarization to chemical stimuli [61, 62]. In DRG neurones isolated from adult rats, PGE$_2$ has a similar sensitizing action on responses to depolarizing stimuli but, in addition, this agent directly depolarizes some neurones (up to 10%) [63].

In small diameter sensory neurones displaying the TTX-resistant Na$^+$ current attributed to Na$_V$1.8, application of PGE$_2$ produces an increase in the amplitude of this current, a hyperpolarizing shift in its activation curve, and an increase in its rates of activation and inactivation [64, 65]. Similar changes have also been reported following application of the hyperalgesic agents, 5-hydroxytyrptamine, adenosine and NGF [65, 66]. In sensory nerve terminals, the hyperpolarizing shift in the activation curve and the increased rate of activation would be expected to reduce the voltage threshold for initiating action potentials. In addition, the increased rate of inactivation should speed action potential repolarization and allow higher frequencies of action potential discharge. Together, these changes would be expected to both reduce the sensory threshold and increase the response to a fixed stimulus.

Activators of protein kinase A (PKA) mimic the effects of PGE$_2$ on the TTX-resistant Na$^+$ current and blockers of PKA inhibit the actions of PGE$_2$ on this current [64, 67]. These findings indicate that PGE_2 modulation of the TTX-resistant Na^+ current is mediated, at least in part, through the cyclic AMP-PKA cascade. Blockers of protein kinase C (PKC) also inhibit the actions of PGE_2 on the TTX-resistant Na^+ current but activators of PKC only increase the amplitude of this current [67]. To explain these observations, it has been proposed that PKC activity is necessary to enable PKA modulation of the TTX-resistant Na^+ current [67].

Supporting a role for $Na_V1.8$ Na^+ channels in nociceptor sensitization is the demonstration that inhibiting their synthesis, by intrathecal administration of an antisense deoxynucleotide, produced both an ~50% reduction in the TTX-resistant Na^+ current in small diameter sensory neurones isolated *in vitro* and a reduction in PGE_2-induced mechanical hyperalgesia *in vivo* [68].

In small diameter sensory neurones from $Na_V1.8$ null mice, activating G-proteins with intracellularly applied GTP or GTP-γ-S increases the TTX-resistant persistent Na^+ current attributed to $Na_V1.9$ [69]. This change occurs without substantial alteration in either the voltage dependence or kinetics of this Na^+ current. Upregulation of this Na^+ current causes membrane depolarization and, when neurones are depolarized from a holding potential of −90 mV, reduces both the voltage threshold for eliciting action potentials and accommodation during maintained depolarization. Therefore modulation of this Na^+ current by G-protein-coupled receptors located at the nerve terminal could potentially produce a generalized reduction in sensory thresholds.

Modulation of Na^+ channel expression

In addition to the effects of inflammatory mediators on the behaviour of voltage-gated Na^+ channels, changes in Na^+ channel expression in the cell bodies of neurones projecting to the inflamed tissue have also been reported. Four days following injection of carrageen into the rat hind paw increases in mRNA and protein expression for $Na_V1.3$, $Na_V1.7$ and $Na_V1.8$ were observed in small diameter neurones located within the ipsilateral (carrageen injected) but not the contralateral (saline injected) L4/L5 DRG [70, 71]. $Na_V1.3$ is a TTX-sensitive Na^+ channel that is highly expressed in embryonic sensory neurones but is normally only expressed at low levels in adult sensory neurones. No changes in the expression of $Na_V1.6$ or $Na_V1.9$ were detected. The changes in Na^+ channel expression were associated with a parallel increase in the amplitude of both TTX-sensitive and TTX-resistant Na^+ currents recorded in small diameter neurones isolated from the ipsilateral L4/L5 ganglia [70, 71]. The voltage-dependence of activation of the Na^+ currents did not differ between neurones isolated from the ipsilateral and contralateral ganglia. Inflammation produced by injection of Freund's complete adjuvant (FCA) into the rat hind paw also increases the expression of $Na_V1.7$ and $Na_V1.8$ protein in DRG neurones supplying the hind limb [72, 73].

An elevation in NGF in the inflamed tissue [74] has been suggested to provide a maintained trophic influence that induces upregulation of Na$^+$ channel expression in the cell bodies of sensory neurones [70, 71]. In accord with this idea, NGF increases the expression of Na$_V$1.8 in cultured DRG neurones [75]. In addition, injection of NGF into the rat hind paw increased expression of Na$_V$1.7 protein in DRG neurones supplying the hind limb, an effect that was reduced when the NGF was sequestered with anti-NGF [76]. In other neuronal cell types NGF upregulates the expression of TTX-sensitive Na$^+$ channels (see [70]). Recently it has been reported that pretreatment with the cyclooxygenase inhibitors, ibuprofen (non-selective) or NS-398 (COX-2 selective), inhibits upregulation of Na$_V$1.7 and Na$_V$1.8 expression following injection of FCA into the rat hind paw [73]. This finding indicates that products of the cyclooxygenase pathway also play a role in triggering increased expression of these Na$^+$ channel subunits.

An increase in the density of both TTX-sensitive and TTX-resistant Na$^+$ channels in sensory nerve terminals would be expected to reduce the voltage threshold for initiating action potentials. As both Na$_V$1.3 and Na$_V$1.7 channels generate an inward current during slow ramp depolarizations that activates near the resting membrane potential [25, 77], increased expression of these channels could potentially play an important role in boosting stimulus-evoked depolarizations. In addition, as Na$_V$1.3 recovers from inactivation very rapidly, increased expression of this α-subunit may enable the nerve terminals to generate action potentials at higher frequencies [77].

Conclusion

Both C and Aδ nociceptive neurones express multiple subtypes of Na$^+$ channel. In particular, TTX-resistant Na$^+$ channels appear to play an important role in determining the behaviour of these neurones. Present evidence suggests that initiation of nerve impulses in the sensory nerve terminals of nociceptors is dependent on the activation of TTX-resistant channels. Furthermore, the voltage dependence and kinetics of these Na$^+$ channels may, at least in part, explain why these receptors have high sensory thresholds. In accord with this idea, a range of hyperalgesic agents released in inflamed tissue have been demonstrated to modify the behaviour of Na$^+$ current attributed to Na$_V$1.8 channels in a manner that will lower the voltage threshold for initiating action potentials. However, this action of hyperalgesic agents is likely to be only one of a range of mechanisms that contribute to inflammatory pain. For example, increased expression of both TTX-sensitive and TTX-resistant Na$^+$ channels may also contribute to lowering the voltage threshold and changing the firing characteristics of nociceptive nerve endings.

Based on current evidence, selective blockade of Na$^+$ channel subtypes, in particular Na$_V$1.8 and/or Na$_V$1.9, may prove useful in controlling painful signals arising from damaged tissue without interfering with other neural functions. Blockade of these channel subtypes may also prove useful in inhibiting neurogenic inflammation of local origin.

References

1   Campbell JN, Meyer RA (1996) Cutaneous nociceptors. In: C Belmonte, F Cervero (eds): *Neurobiology of nociceptors*. Oxford University Press, Oxford, 119–141

2   Black JA, Dib-Hajj S, McNabola K, Jeste S, Rizzo MA, Kocsis JD, Waxman SG (1996) Spinal sensory neurons express multiple sodium channel α-subunit mRNAs. *Brain Res Mol Brain Res* 43: 117–131

3   Sangameswaran L, Delgado SG, Fish LM, Koch BD, Jakeman LB, Stewart GR, Sze P, Hunter JC, Eglen RM, Herman RC (1996) Structure and function of a novel voltage-gated, tetrodotoxin-resistant sodium channel specific to sensory neurons. *J Biol Chem* 271: 5953–5956

4   Akopian AN, Sivilotti L, Wood JN (1996) A tetrodotoxin-resistant voltage-gated sodium channel expressed by sensory neurons. *Nature* 379: 257–262

5   Dib-Hajj SD, Tyrrell L, Black JA, Waxman SG (1998) NaN, a novel voltage-gated Na channel, is expressed preferentially in peripheral sensory neurons and down-regulated after axotomy. *Proc Natl Acad Sci USA* 95: 8963–8968

6   Shah BS, Stevens EB, Gonzalez MI, Bramwell S, Pinnock RD, Lee K, Dixon AK (2000) β3, a novel auxiliary subunit for the voltage-gated sodium channel, is expressed preferentially in sensory neurons and is upregulated in the chronic constriction injury model of neuropathic pain. *Eur J Neurosci* 12: 3985–3990

7   Takahashi N, Kikuchi S, Dai Y, Kobayashi K, Fukuoka T, Noguchi K (2003) Expression of auxiliary β subunits of sodium channels in primary afferent neurons and the effect of nerve injury. *Neuroscience* 121: 441–450

8   Amaya F, Decosterd I, Samad TA, Plumpton C, Tate S, Mannion RJ, Costigan M, Woolf CJ (2000) Diversity of expression of the sensory neuron-specific TTX-resistant voltage-gated sodium ion channels SNS and SNS2. *Mol Cell Neurosci* 15: 331–342

9   Benn SC, Costigan M, Tate S, Fitzgerald M, Woolf CJ (2001) Developmental expression of the TTX-resistant voltage-gated sodium channels Na$_V$1.8 (SNS) and Na$_V$1.9 (SNS2) in primary sensory neurons. *J Neurosci* 21: 6077–6085

10   Caterina MJ, Rosen TA, Tominaga M, Brake AJ, Julius D (1999) A capsaicin-receptor homologue with a high threshold for noxious heat. *Nature* 398: 436–441

11   Porreca F, Lai J, Bian D, Wegert S, Ossipov MH, Eglen RM, Kassotakis L, Novakovic S, Rabert DK, Sangameswaran L et al (1999) A comparison of the potential role of the tetrodotoxin-insensitive sodium channels, PN3/SNS and NaN/SNS2, in rat models of chronic pain. *Proc Natl Acad Sci USA* 96: 7640–7644

12   Tzoumaka E, Tischler AC, Sangameswaran L, Eglen RM, Hunter JC, Novakovic SD

(2000) Differential distribution of the tetrodotoxin-sensitive rPN4/NaCh6/Scn8a sodium channel in the nervous system. *J Neurosci* Res 60: 37–44

13   Djouhri L, Fang X, Okuse K, Wood JN, Berry CM, Lawson SN (2003) The TTX-resistant sodium channel Na$_V$1.8 (SNS/PN3): expression and correlation with membrane properties in rat nociceptive primary afferent neurons. *J Physiol* 550: 739–752

14   Fang X, Djouhri L, Black JA, Dib-Hajj SD, Waxman SG, Lawson SN (2002) The presence and role of the tetrodotoxin-resistant sodium channel Na$_V$1.9 (NaN) in nociceptive primary afferent neurons. *J Neurosci* 22: 7425–7433

15   Djouhri L, Newton R, Levinson SR, Berry CM, Carruthers B, Lawson SN (2003) Sensory and electrophysiological properties of guinea-pig sensory neurones expressing Na$_V$ 1.7 (PN1) Na$^+$ channel α-subunit protein. *J Physiol* 546: 565–576

16   Caldwell JH, Schaller KL, Lasher RS, Peles E, Levinson SR (2000) Sodium channel Na$_V$1.6 is localized at nodes of Ranvier, dendrites, and synapses. *Proc Natl Acad Sci USA* 97: 5616–5620

17   Black JA, Renganathan M, Waxman SG (2002) Sodium channel Na$_V$1.6 is expressed along nonmyelinated axons and it contributes to conduction. *Brain Res Mol Brain Res* 105: 19–28

18   Black JA, Waxman SG (2002) Molecular identities of two tetrodotoxin-resistant sodium channels in corneal axons. *Exp Eye Res* 75: 193–199

19   Fjell J, Hjelmstrom P, Hormuzdiar W, Milenkovic M, Aglieco F, Tyrrell L, Dib-Hajj S, Waxman SG, Black JA (2000) Localization of the tetrodotoxin-resistant sodium channel NaN in nociceptors. *Neuroreport* 11: 199–202

20   Vulchanova L, Olson TH, Stone LS, Riedl MS, Elde R, Honda CN (2001) Cytotoxic targeting of isolectin IB4-binding sensory neurons. *Neuroscience* 108: 143–155

21   Coward K, Aitken A, Powell A, Plumpton C, Birch R, Tate S, Bountra C, Anand P (2001) Plasticity of TTX-sensitive sodium channels PN1 and brain III in injured human nerves. *Neuroreport* 12: 495–500

22   Coward K, Plumpton C, Facer P, Birch R, Carlstedt T, Tate S, Bountra C, Anand P (2000) Immunolocalization of SNS/PN3 and NaN/SNS2 sodium channels in human pain states. *Pain* 85: 41–50

23   Dietrich PS, McGivern JG, Delgado SG, Koch BD, Eglen RM, Hunter JC, Sangameswaran L (1998) Functional analysis of a voltage-gated sodium channel and its splice variant from rat dorsal root ganglia. *J Neurochem* 70: 2262–2272

24   Klugbauer N, Lacinova L, Flockerzi V, Hofmann F (1995) Structure and functional expression of a new member of the tetrodotoxin-sensitive voltage-activated sodium channel family from human neuroendocrine cells. *Embo J* 14: 1084–1090

25   Cummins TR, Howe JR, Waxman SG (1998) Slow closed-state inactivation: a novel mechanism underlying ramp currents in cells expressing the hNE/PN1 sodium channel. *J Neurosci* 18: 9607–9619

26   Sangameswaran L, Fish LM, Koch BD, Rabert DK, Delgado SG, Ilnicka M, Jakeman LB, Novakovic S, Wong K, Sze P et al (1997) A novel tetrodotoxin-sensitive, voltage-

gated sodium channel expressed in rat and human dorsal root ganglia. *J Biol Chem* 272: 14805–14809

27   Dib-Hajj S, Black JA, Cummins TR, Waxman SG (2002) NaN/ $Na_V1.9$: a sodium channel with unique properties. *Trends Neurosci* 25: 253–259

28   Ogata N, Tatebayashi H (1993) Kinetic analysis of two types of Na^+ channels in rat dorsal root ganglia. *J Physiol* 466: 9–37

29   Elliott AA, Elliott JR (1993) Characterization of TTX-sensitive and TTX-resistant sodium currents in small cells from adult rat dorsal root ganglia. *J Physiol* 463: 39–56

30   Rush AM, Brau ME, Elliott AA, Elliott JR (1998) Electrophysiological properties of sodium current subtypes in small cells from adult rat dorsal root ganglia. *J Physiol* 511: 771–789

31   Akopian AN, Souslova V, England S, Okuse K, Ogata N, Ure J, Smith A, Kerr BJ, McMahon SB, Boyce S et al (1999) The tetrodotoxin-resistant sodium channel SNS has a specialized function in pain pathways. *Nat Neurosci* 2: 541–548

32   Cummins TR, Dib-Hajj SD, Black JA, Akopian AN, Wood JN, Waxman SG (1999) A novel persistent tetrodotoxin-resistant sodium current in SNS-null and wild-type small primary sensory neurons. *J Neurosci* 19: RC43

33   Herzog RI, Cummins TR, Waxman SG (2001) Persistent TTX-resistant Na^+ current affects resting potential and response to depolarization in simulated spinal sensory neurons. *J Neurophysiol* 86: 1351–1364

34   Yoshida S, Matsuda Y (1979) Studies on sensory neurons of the mouse with intracellular-recording and horseradish peroxidase-injection techniques. *J Neurophysiol* 42: 1134–1145

35   Lopez de Armentia M, Cabanes C, Belmonte C (2000) Electrophysiological properties of identified trigeminal ganglion neurons innervating the cornea of the mouse. *Neuroscience* 101: 1109–1115

36   Villiere V, McLachlan EM (1996) Electrophysiological properties of neurons in intact rat dorsal root ganglia classified by conduction velocity and action potential duration. *J Neurophysiol* 76: 1924–1941

37   Ritter AM, Mendell LM (1992) Somal membrane properties of physiologically identified sensory neurons in the rat: effects of nerve growth factor. *J Neurophysiol* 68: 2033–2041

38   Quasthoff S, Grosskreutz J, Schroder JM, Schneider U, Grafe P (1995) Calcium potentials and tetrodotoxin-resistant sodium potentials in unmyelinated C fibres of biopsied human sural nerve. *Neuroscience* 69: 955–965

39   Novakovic SD, Tzoumaka E, McGivern JG, Haraguchi M, Sangameswaran L, Gogas KR, Eglen RM, Hunter JC (1998) Distribution of the tetrodotoxin-resistant sodium channel PN3 in rat sensory neurons in normal and neuropathic conditions. *J Neurosci* 18: 2174–2187

40   Brock JA, McLachlan EM, Belmonte C (1998) Tetrodotoxin-resistant impulses in single nociceptor nerve terminals in guinea-pig cornea. *J Physiol* 512: 211–217

41   Belmonte C, Garcia-Hirschfeld J, Gallar J (1997) Neurobiology of ocular pain. *Progress in Retinal and Eye Research* 16: 117–156

42   Belmonte C, Acosta MC, Gallar J (2004) Neural basis of sensation in intact and injured corneas. *Exp Eye Res* 78: 513–525

43   Acosta MC, Belmonte C, Gallar J (2001) Sensory experiences in humans and single-unit activity in cats evoked by polymodal stimulation of the cornea. *J Physiol* 534: 511–525

44   Roy ML, Narahashi T (1992) Differential properties of tetrodotoxin-sensitive and tetrodotoxin-resistant sodium channels in rat dorsal root ganglion neurons. *J Neurosci* 12: 2104–2111

45   Strassman AM, Raymond SA (1999) Electrophysiological evidence for tetrodotoxin-resistant sodium channels in slowly conducting dural sensory fibers. *J Neurophysiol* 81: 413–424

46   Kirchhoff CG, Reeh PW, Waddell PJ (1986) Sensory endings of C- and A-fibres are differentially sensitive to tetrodotoxin in rat skin *in vitro*. *J Physiol* 418: P116

47   Brock JA, Pianova S, Belmonte C (2001) Differences between nerve terminal impulses of polymodal nociceptors and cold sensory receptors of the guinea-pig cornea. *J Physiol* 533: 493–501

48   Carr RW, Pianova S, Brock JA (2002) The effects of polarizing current on nerve terminal impulses recorded from polymodal and cold receptors in the guinea-pig cornea. *J Gen Physiol* 120: 395–405

49   Smith DO (1988) Determinants of nerve terminal excitability. In: PW Lanfield, SA Deadwyler (eds): *Neurology and Neurobiology*, vol. 35, Long-term potentiation. Alan Liss Inc., New York, 411–438

50   Schild JH, Kunze DL (1997) Experimental and modeling study of Na^+ current heterogeneity in rat nodose neurons and its impact on neuronal discharge. *J Neurophysiol* 78: 3198–3209

51   Blair NT, Bean BP (2002) Roles of tetrodotoxin (TTX)-sensitive Na^+ current, TTX-resistant Na^+ current, and Ca^{2+} current in the action potentials of nociceptive sensory neurons. *J Neurosci* 22: 10277–10290

52   Caffrey JM, Eng DL, Black JA, Waxman SG, Kocsis JD (1992) Three types of sodium channels in adult rat dorsal root ganglion neurons. *Brain Res* 592: 283–297

53   Renganathan M, Cummins TR, Waxman SG (2001) Contribution of $Na_V1.8$ sodium channels to action potential electrogenesis in DRG neurons. *J Neurophysiol* 86: 629–640

54   Schaible HG, Schmidt RF (1988) Excitation and sensitization of fine articular afferents from cat's knee joint by prostaglandin E_2. *J Physiol* 403: 91–104

55   Birrell GJ, McQueen DS, Iggo A, Coleman RA, Grubb BD (1991) PGI_2-induced activation and sensitization of articular mechanonociceptors. *Neurosci Lett* 124: 5–8

56   Mizumura K, Sato J, Kumazawa T (1987) Effects of prostaglandins and other putative chemical intermediaries on the activity of canine testicular polymodal receptors studied *in vitro*. *Pflugers Arch* 408: 565–572

57   Mizumura K, Minagawa M, Tsujii Y, Kumazawa T (1993) Prostaglandin E_2-induced

sensitization of the heat response of canine visceral polymodal receptors *in vitro*. *Neurosci Lett* 161: 117–119

58   Khasar SG, Green PG, Levine JD (1993) Comparison of intradermal and subcutaneous hyperalgesic effects of inflammatory mediators in the rat. *Neurosci Lett* 153: 215–218

59   Schuligoi R, Donnerer J, Amann R (1994) Bradykinin-induced sensitization of afferent neurons in the rat paw. *Neuroscience* 59: 211–215

60   Southall MD, Vasko MR (2001) Prostaglandin receptor subtypes, EP3C and EP4, mediate the prostaglandin E_2-induced cAMP production and sensitization of sensory neurons. *J Biol Chem* 276: 16083–16091

61   Nicol GD, Cui M (1994) Enhancement by prostaglandin E_2 of bradykinin activation of embryonic rat sensory neurones. *J Physiol* 480: 485–492

62   Cui M, Nicol GD (1995) Cyclic AMP mediates the prostaglandin E_2-induced potentiation of bradykinin excitation in rat sensory neurons. *Neuroscience* 66: 459–466

63   Kasai M, Mizumura K (2001) Effects of PGE_2 on neurons from rat dorsal root ganglia in intact and adjuvant-inflamed rats: role of NGF on PGE_2-induced depolarization. *Neurosci Res* 41: 345–353

64   England S, Bevan S, Docherty RJ (1996) PGE_2 modulates the tetrodotoxin-resistant sodium current in neonatal rat dorsal root ganglion neurones via the cyclic AMP-protein kinase A cascade. *J Physiol* 495: 429–440

65   Gold MS, Reichling DB, Shuster MJ, Levine JD (1996) Hyperalgesic agents increase a tetrodotoxin-resistant Na^+ current in nociceptors. *Proc Natl Acad Sci USA* 93: 1108–1112

66   Zhang YH, Vasko MR, Nicol GD (2002) Ceramide, a putative second messenger for nerve growth factor, modulates the TTX-resistant Na^+ current and delayed rectifier K^+ current in rat sensory neurons. *J Physiol* 544: 385–402

67   Gold MS, Levine JD, Correa AM (1998) Modulation of TTX-R INa by PKC and PKA and their role in PGE_2-induced sensitization of rat sensory neurons *in vitro*. *J Neurosci* 18: 10345–10355

68   Khasar SG, Gold MS, Levine JD (1998) A tetrodotoxin-resistant sodium current mediates inflammatory pain in the rat. *Neurosci Lett* 256: 17–20

69   Baker MD, Chandra SY, Ding Y, Waxman SG, Wood JN (2003) GTP-induced tetrodotoxin-resistant Na^+ current regulates excitability in mouse and rat small diameter sensory neurones. *J Physiol* 548: 373–382

70   Black JA, Liu S, Tanaka M, Cummins TR, Waxman SG (2004) Changes in the expression of tetrodotoxin-sensitive sodium channels within dorsal root ganglia neurons in inflammatory pain. *Pain* 108: 237–247

71   Tanaka M, Cummins TR, Ishikawa K, Dib-Hajj SD, Black JA, Waxman SG (1998) SNS Na^+ channel expression increases in dorsal root ganglion neurons in the carrageenan inflammatory pain model. *Neuroreport* 9: 967–972

72   England JD, Gould HJ, Liu AG, Koszowski SR, Levinson SR (1998) Inflammation induces a rapid upregulation of the novel sodium channel, PN1, in sensory neurones. *Neurology* 47 (Suppl 4): A184

73 Gould HJ, England JD, Soignier RD, Nolan P, Minor LD, Liu ZP, Levinson SR, Paul D (2004) Ibuprofen blocks changes in Na_V 1.7 and 1.8 sodium channels associated with complete Freund's adjuvant-induced inflammation in rat. *J Pain* 5: 270–280

74 Woolf CJ, Safieh-Garabedian B, Ma QP, Crilly P, Winter J (1994) Nerve growth factor contributes to the generation of inflammatory sensory hypersensitivity. *Neuroscience* 62: 327–331

75 Black JA, Langworthy K, Hinson AW, Dib-Hajj SD, Waxman SG (1997) NGF has opposing effects on Na^+ channel III and SNS gene expression in spinal sensory neurons. *Neuroreport* 8: 2331–2335

76 Gould HJ, Gould TN, England JD, Paul D, Liu ZP, Levinson SR (2000) A possible role for nerve growth factor in the augmentation of sodium channels in models of chronic pain. *Brain Res* 854: 19–29

77 Cummins TR, Aglieco F, Renganathan M, Herzog RI, Dib-Hajj SD, Waxman SG (2001) Na_V1.3 sodium channels: rapid repriming and slow closed-state inactivation display quantitative differences after expression in a mammalian cell line and in spinal sensory neurons. *J Neurosci* 21: 5952–5961

Signalling cascades that modulate the activity of sodium channels in sensory neurons

Grant D. Nicol

Department of Pharmacology and Toxicology, Indiana University School of Medicine, 635 Barnhill Drive, Indianapolis, IN 46202, USA

Introduction

It is well established that inflammatory mediators can heighten the sensitivity to a variety of different modalities of sensory stimulation. Early work demonstrated that inflammatory mediators, such as prostaglandin E_2, serotonin, or nerve growth factor, lower the threshold to nociceptive stimuli in animal models of pain. Later work has shown that a large part of this enhanced sensitivity results directly from the altered sensitivity or excitability of the sensory neurons themselves. All of these inflammatory mediators are known to act via membrane receptors, and it comes as no surprise that nociceptive sensory neurons express many of these receptors. Ligand binding to these receptors results in the activation of downstream signalling cascades which can ultimately regulate or modulate the activity of ion channels that are critical in setting the state of excitability in sensory neurons. This review will focus on the signalling cascades that modulate the activity of voltage-dependent sodium channels that give rise to the augmented sensitivity to various kinds of stimuli. This alteration in the sensitivity or threshold will be referred to as sensitization. In addition to modulation of channel activity, another important mechanism that can modify the state of excitability is the transcriptional change that leads to alterations in the levels of expression for different ion channels. Several recent reviews have discussed the changes in sodium channel expression after different types of nerve injury, such as that arising after neuropathic or inflammatory pain states ([1–9] and Chapter by J.J. Clare, this volume), therefore such observations regarding sodium channels will not be discussed.

Activation of the protein kinase A pathway

In early behavioral measurements, Ferreira and Nakamura [10] originally demonstrated that activation of the cyclic AMP pathway might be involved in the enhancement of sensitivity to noxious mechanical stimulation. This was based on their observation that the hyperalgesia produced by injection of the proinflammatory prostaglandin, prostaglandin E_2 (PGE$_2$), into the paw was exactly paralleled by a membrane permeant analog of cyclic AMP, dibutyryl cyclic AMP. This work was later confirmed by Levine's group [11]. Biochemical studies have very clearly established that elevations in intracellular cyclic AMP result from receptor activation that is coupled to the stimulatory G protein, Gs, which then activates adenylyl cyclase. Elevated levels of cyclic AMP activate protein kinase A (PKA); this kinase can then phosphorylate many different substrate proteins, one important group being ion channels.

This idea that a receptor-mediated elevation in cyclic AMP level played a key role in the sensitization of sensory neurons was clearly elucidated in two important studies by Gold et al. [12] and England et al. [13]. Exposure to PGE$_2$ augmented the tetrodotoxin-resistant sodium current (TTX-R I_{Na}) by 10–20% with a leftward (more hyperpolarized) shift in the voltage dependence for activation of this current by about 5 mV. In addition to PGE$_2$, other inflammatory agents such as adenosine and serotonin [12] as well as cyclic AMP analogs and the adenylyl cyclase activator, forskolin, [13, 14] exhibited similar enhancements of TTX-R I_{Na}. The effects of PGE$_2$ on the inactivation properties of TTX-R I_{Na} are not clear since England et al. report a −5 mV shift in the half-inactivation voltage whereas Gold et al. observed no change. Surprisingly, exposure to PGE$_2$ failed to increase the amplitude of the TTX-sensitive I_{Na} (TTX-S I_{Na}) ([12] and see below). The notion that this alteration in the TTX-R I_{Na} resulted from a cyclic AMP-dependent phosphorylation of the channel was indicated by the observation that internal perfusion with a peptide inhibitor of PKA blocked the capacity of PGE$_2$ to enhance the current [13].

The critical role of PKA-mediated phosphorylation of the sodium channel was demonstrated by Fitzgerald et al. [15], wherein they mutated the five serine residues (consensus PKA phosphorylation sites) to alanines in the intracellular loop between transmembrane domains I and II in the SNS/PN3 sodium channel (now known as Na$_V$1.8). The cDNA for Na$_V$1.8 was expressed in COS-7 cells and exhibited properties that were similar to those described in native DRG neurons. The currents conducted by Na$_V$1.8 were enhanced after exposure to either forskolin or 8-Br cyclic AMP, and the half-activation voltage was shifted to more hyperpolarized values (~8 mV). Although the currents conducted by the mutant channel exhibited properties similar to the wild-type Na$_V$1.8, forskolin or 8-Br cyclic AMP failed to augment the current or shift the voltage dependence of activation. These results clearly demonstrated that PKA-mediated phosphorylation of Na$_V$1.8 played a causal role in the augmentation of the peak TTX-R I_{Na} as well as the shift to more hyperpolarized voltages.

Serotonin or 5-HT is known to be a potent proinflammatory mediator [16–19] and also produces sensitization of behavioral responses in animal models of pain [20–22]. The hyperalgesic response to 5-HT was demonstrated to involve the cyclic

AMP signalling pathway [23, 24]. Analogous to PGE_2, studies in isolated small-diameter sensory neurons showed that 5-HT enhanced the TTX-R I_{Na} [12, 25]. Using a neuronal characterization scheme developed in Scroggs' laboratory [26], 5-HT appeared to sensitize TTX-R I_{Na} in only type 2 neurons (small diameter, capsaicin-sensitive, long duration action potential with a "hump") whereas PGE_2 was effective in augmenting I_{Na} in all four neuronal subtypes (~90, 30, 20, and 13% enhancements in types 1, 2, 3 and 4, respectively). However, PGE_2 modulated only TTX-R I_{Na} in types 1 and 2, whereas PGE_2 was effective on TTX-S I_{Na} in types 3 and 4. Later work demonstrated that the effects of 5-HT were mediated by the cyclic AMP pathway in these type 2 sensory neurons [27]. Exposure of type 2 neurons to the phosphodiesterase inhibitor, IBMX, augmented TTX-R I_{Na} suggesting that under normal conditions there is some basal level of adenylyl cyclase/PKA activity. In contrast to the effects of PGE_2, 5-HT augments the peak current without shifting the voltage dependency for either activation or inactivation [28]. These results are interesting in that they suggest that inflammatory mediators such as PGE_2 and 5-HT may have overlapping actions in only some types of sensory neurons whereas their actions may be unique in other subtypes.

Activation of the protein kinase C pathway

Agonists, such as bradykinin, bind to G protein-coupled receptors that consequently lead to the release of IP3 (producing an elevation in intracellular Ca^{2+}) and diacylglycerol. Release of Ca^{2+} and liberation of diacylglycerol results in the activation of conventional subtypes of protein kinase C (PKC). Early studies using phorbol esters to directly activate PKC showed that this pathway was involved in the stimulation of primary afferents [29, 30] as well as isolated sensory neurons [31]. Later work by Schepelmann et al. [32] demonstrated that phorbol esters could directly excite primary afferents innervating the knee joint, but also, phorbol esters lead to a sensitization of the response to passive movement of the joint. Consistent with these observations, Barber and Vasko [33] found that low concentrations of the phorbol ester, PDBu, enhanced the release of neuropeptides from isolated sensory neurons. Taken together, these observations suggested that activation of PKC played an important role in augmenting the neuronal sensitivity to stimulation. Indeed, studies in other neuronal systems have demonstrated that PKC has the capacity to modulate the activity of a variety of ion channels [34–37] and could therefore account for the enhanced sensitivity.

Analogous to the actions of PGE_2, stimulation of PKC by the phorbol esters, PMA or PDBu, enhanced the TTX-R I_{Na} by about 25–35% in small-diameter sensory neurons isolated from adult rats [14]. The phorbol ester-induced increase was blocked by pretreatment with inhibitors of PKC. Interestingly, treatment with PKC inhibitors alone reduced TTX-R I_{Na} by ~50%, whereas a PKA inhibitor had little effect; these findings suggest that there is ongoing PKC-mediated phosphorylation of this channel under basal conditions. The lack of effect by the PKA inhibitor is in contrast to the IBMX results of Cardenas et al. [27], unless basal activity of adenylyl cyclase/PKA is unique to type 2 sensory neurons. Unlike PGE_2, phorbol esters did not alter the voltage dependence for activation of TTX-R I_{Na}.

It is well documented that different signalling pathways interact with one another or influence each other's activity, i.e., cross-talk [38–41]. Indeed, there appears to be an interaction between the PKA and PKC pathways in modulating TTX-R I_{Na} [14]. Pretreatment of the neurons with a PKC inhibitor (internally perfused PKC_{19-36} or bath-applied staurosporine) significantly reduced the ability of forskolin to enhance TTX-R I_{Na}, whereas pretreatment with inhibitors of PKA (internally perfused WIPTIDE or Rp-cAMPS) failed to alter the phorbol ester-induced increase in TTX-R I_{Na}. Similarly, PKC inhibitors attenuated the sensitizing actions of PGE_2. These results suggest that the phosphorylation mediated by PKC is permissive for the modulation of TTX-R I_{Na} by the cyclic AMP pathway. These results are consistent with earlier studies by Catterall's group that demonstrated PKC-induced phosphorylation of the serine residue at position 1506 (located in the intracellular loop between domains III and IV and is believed to regulate inactivation of the channel) in the TTX-S channel ($Na_V1.2$) was required before the serines between domains I and II could be phosphorylated by PKA (see review by [42]; see discussion in [14]). Levine's group has reported observations that are consistent with this idea. Khasar et al. [43] showed that application of epinephrine, which can induce mechanical and thermal hyperalgesia, augments the TTX-R I_{Na} in a manner similar to PGE_2 (~40% increase in the peak current with a 10 mV hyperpolarizing shift in the half-activation voltage). This effect was prevented by the PKA inhibitor Rp-cAMPS indicating that this sensitization was mediated by epinephrine's activation of the cyclic AMP/PKA pathway. However, pretreatment with the PKC inhibitor, bisindoylmaleimide, attenuated the epinephrine-induced sensitization of TTX-R I_{Na} by about 50%. Thus, these findings indicate that there are important interactions between the PKC and PKA signalling pathways that regulate the activity of the channel(s) conducting TTX-R I_{Na}. The cellular mechanisms that result in this dual PKC/PKA modulation of TTX-R I_{Na} remain unknown.

Activation of the PKC signalling pathway and its modulation of TTX-R I_{Na} may play an important role in the pain causing actions of endothelin-1, a potent vasoconstrictor ([44], see recent work of G. Davar's laboratory). The pain-inducing effects of endothelin-1 are known to be mediated by the ET_A receptor subtype; this is a G protein-coupled receptor linked to activation of PKC [45–48]. Recent work by Zhou et al. [49] demonstrated that the behavioral sensitization produced by endothelin-1 may, in part, result from the modulation of the gating of TTX-R I_{Na}. In approximately half those neurons (10/17) exhibiting TTX-R I_{Na}, endothelin-1 produced a hyperpolarizing shift in the half-activation voltage that corresponded to ~8 mV. Although the voltage dependence for activation was modified, the maximal conductance was not altered. Endothelin-1 had no effect on the gating of TTX-S I_{Na} in larger sensory neurons (which should exhibit little if any TTX-R I_{Na}; the presence or absence of ET_A in these neurons was not examined). This suggests that endothelin-1 modifies the gating of these TTX-R channels but does not change the number of channels that are capable of activation. The effects of endothelin-1 on TTX-R channels differ from previous observations in two ways. First, the unchanged conductance is different from the sensitization produced by PGE_2 as well as the effects of phorbol esters reported by Gold et al. [14]. Second, if the endothelin-1-induced enhancement is mediated by activation of PKC, then the hyperpolarizing shift in the half-activation voltage is different than described for phorbol esters [24]. These findings raise a curious question, in that, activation of presumably the same signalling cascade (i.e., PGE_2 and cyclic AMP or endothelial-1/phorbol esters and PKC) in some studies modifies the voltage dependency whereas in others it does not. Does this result from real physiological differences in the regulation of channel activity or is it dependent on the methodologies unique to each laboratory?

Sensitization produced by nerve growth factor

It is well established that the levels of nerve growth factor (NGF) are elevated in inflammatory exudates [50] and that exposure to NGF produces a hyperalgesic response in animal models. Early work by Mendell's group showed that injection of NGF into the hind paw of a rat produced a rapid onset of thermal hyperalgesia (tens of minutes) and a much more delayed beginning of mechanical hyperalgesia (several hours) [51]. These sensitizing actions of NGF appear to be directly on the sensory nerve since NGF increased the firing frequency of the isolated saphenous nerve in response to thermal stimulation [52]. There is a large body of work that has described the trophic actions of NGF on the expression levels of sodium channels in a variety of model systems [53–57], however, there have been few studies that have examined the acute modulatory actions of NGF on the properties of sodium channels. Recently Zhang et al. [58] observed that exposure to NGF caused an increase in the number of action potentials evoked by a ramp of depolarizing current in small diameter rat sensory neurons that were sensitive to capsaicin. This sensitization was due, in part, to a rapid enhancement (< 2 min) of the peak TTX-R I_{Na} (see Fig. 1). Associated with this increased current was a hyperpolarizing shift of ~6 mV in the half-activation voltage for TTX-R I_{Na}. The shift in activation voltage is similar to those reported for the effects of PGE_2 and endothelin-1 (see above). Surprisingly, this study found that the NGF-induced enhancement was mediated by activation of the p75 neurotrophin receptor. The p75 receptor is coupled to the sphingomyelin pathway and the liberation of ceramide [59, 60]. Activation of the sphingomyelin pathway was indicated by several observations. First, internally perfused sphingomyelinase (the enzyme that liberates ceramide from sphingomyelin) increased the

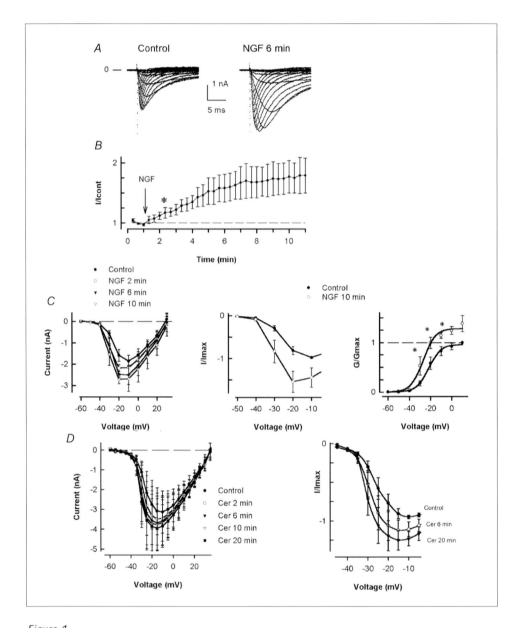

Figure 1

NGF and ceramide enhance the TTX-RI$_{Na}$ in adult sensory neurons

A: The effects of 100 ng ml^{-1} NGF on representative current traces under control conditions (left) compared to those after a 6 min exposure to NGF (right). The line labelled zero represents the zero current value. B: the time course of NGF's action. The peak TTX-R I$_{Na}$, was obtained for a voltage step from –60 to –20 mV, and this step was repeated every 20 s. NGF

number of action potentials in a manner very similar to NGF. Second, inhibition of the neutral form of sphingomyelinase by glutathione [61, 62] prevented the sensitizing actions of NGF on the evoked action potentials. Third, in the presence of glutathione, exogenous application of a membrane permeable ceramide analog increased the number of evoked action potentials. Fourth, exogenous ceramide augmented TTX-R I_{Na} and was analogous to the actions of NGF. At present, the cellular mechanism(s) whereby ceramide or other downstream mediators of the sphingomyelin cascade enhance TTX-R I_{Na} is not known and awaits further study. Such results would suggest that there are multiple pathways that can modulate TTX-R I_{Na} in sensory neurons and raises important questions as to whether this modulation targets the same phosphorylation site, are there multiple sites each specific to a particular signalling pathway, and do these pathways act in a simple additive manner or does activation of one pathway potentiate or facilitate the actions of a parallel pathway in a synergistic manner?

Although the modulatory actions of NGF on TTX-S I_{Na} in sensory neurons have not been investigated, NGF appears to suppress the TTX-S I_{Na} in both differentiated and undifferentiated PC12 cells [63]. Exposure of PC12 cells to NGF produced ~40% reduction in the peak I_{Na} attaining maximal suppression within ~90 s. The voltage dependence for inactivation was shifted to more hyperpolarized voltages, however, activation was not altered. The NGF-induced decrease was mediated by

---

was added at the indicated time. The asterisk represents the first time point that was significantly different from the control values. The data points represent the average obtained from three neurons. Left panel in C, the current-voltage relations obtained before and after treatment with NGF. Treatment times of 6 and 10 min produced a significant increase in the peak TTX-R I_{Na} for voltage steps between –40 and +20 mV (RM ANOVA). The membrane voltage was held at –60 mV; activation of the currents was determined by voltage steps of 30 ms that were applied at 5 s intervals in +5 or +10 mV increments to +60 mV. Middle panel in C, the normalized current-voltage relation and the effects of NGF. Peak currents were normalized to their respective control values obtained for the step to –10 mV. Significant increases were obtained for voltages between –40 and +10 mV. Right panel in C demonstrates the conductance-voltage relation; data points have been normalized to the conductance obtained at +10 mV. Left panel in D, time-dependent effects of 1 µM ceramide (Cer); the current was increased significantly (RM ANOVA) for voltages between –20 and +15 mV and –25 and +20 mV for 6 min and for both 10 and 20 min exposures, respectively. Right panel in D, the effects of ceramide on the normalized current-voltage relation. Peak currents were normalized to their respective control values obtained for the step to –10 mV. Significant increases were obtained for voltages between –15 and +5 mV and –25 and +20 mV for 10 and 20 min exposures, respectively. Asterisks represent a significant difference (P < 0.05) compared to control.

the TrkA receptor as the receptor tyrosine kinase inhibitor, AG879, blocked the actions of NGF. Activation of other tyrosine kinase receptors, such as epidermal growth factor, also led to the suppression of I_{Na}, suggesting that this may be a general signalling pathway that modulates the levels of sodium channel activity in these cells. Expression of mutant growth factor receptors that lacked the capacity to interact with specific signalling domains suggested that the suppression of I_{Na} depended on the interaction of the receptor with the kinase Src. These results indicate that activation of tyrosine kinase receptors, such as TrkA, in sensory neurons may lead to modulation of TTX-S sodium channels since sensory neurons have many of the signalling pathways resident in PC12 cells. The difficulty in sorting out this question may lie in the isolation of currents conducted by the different subtypes of sodium channels (see below).

Modulation by calmodulin

With increasing appreciation, it is becoming apparent that the activity of many different types of ion channels are modulated by calmodulin. The best examples of this modulation come from studies examining the effects of calmodulin regulation on sodium channels in cardiac myocytes and on the cyclic nucleotide-gated channels found in either visual or olfactory sensory receptors. This literature has been reviewed (see [64–68]). There is a recent report of the modulatory actions of calmodulin on the current conducted by the TTX-S sodium channel $Na_V1.6$ in rat sensory neurons [69]. Calmodulin was demonstrated to interact strongly with GST-fusion proteins for the C-terminal constructs of $Na_V1.2$, $Na_V1.4$, and $Na_V1.6$ for both high and low calcium conditions. This binding was weaker for $Na_V1.1$ and $Na_V1.3$, whereas the interaction with $Na_V1.7$ was only observed under low calcium conditions. Interestingly, none of the C-terminal constructs for the TTX-R sodium channels, $Na_V1.5$, $Na_V1.8$, or $Na_V1.9$, exhibited any interaction with calmodulin [69]. Expression of a TTX-R mutant of $Na_V1.6$ in sensory neurons that were isolated from the dorsal root ganglia of $Na_V1.8$-null mice yielded currents that were very similar to those attributed to TTX-S $Na_V1.6$. Cleverly, the mutation converting the TTX-S $Na_V1.6$ (Y371S) to a TTX-R form permitted the remaining TTX-S $I_{Na}s$ in the $Na_V1.8$-null neurons to be removed by treatment with TTX. Additional mutations in the calmodulin binding domain (the IQ motif) gave peak currents that were reduced between 60–75% of that observed for the TTX-R mutant of $Na_V1.6$ [27]. Overexpression of calmodulin permitted some recovery of the peak current back to "normal" levels and was dependent on the nature of the mutation. The functional significance of this modulation is that when intracellular levels of Ca^{2+} were increased, such as that occurring with a train of action potentials, the interaction of calmodulin with $Na_V1.6$ appeared to slow the rate of inactivation although neither the voltage-dependence for activation nor inactivation were altered. This effect was not observed with a dominant-negative form of calmodulin [69]. These results suggest that changes in intracellular levels of Ca^{2+} via calmodulin may have a role in selectively modulating the extent of TTX-S I_{Na} and therefore its contribution to the overall excitability of the sensory neuron.

Metabotropic GluR suppression of sensitization by PGE_2

Recent work has shown that activation of group II metabotropic glutamate receptors (mGluR) can block both the thermal hyperalgesia as well as the enhancement of Ca^{2+} flux through TRPV1 that is produced by PGE_2 [70]. In this study, the actions of mGluR on the capsaicin-evoked Ca^{2+} response were prevented by pretreatment with pertussis toxin. These results suggest that mGluR activated the inhibitory G protein, Gi, and thereby inhibited the PGE_2-induced activation of adenylyl cyclase. In a similar line of studies, Yang and Gereau [71] demonstrated that activation of group II mGluR by the agonist, ammonium pyrrolidinedithiocarbamate, suppressed the increase in TTX-R I_{Na} that resulted after exposure to forskolin. This suppression was prevented by pretreatment with a group II antagonist, LY341495. Thus, these results suggest that activation of the Gi pathway can block or potentially reverse the sensitization elicited by mediators that increase the levels of cyclic AMP and its consequent activation of PKA.

Unanswered questions

It is most intriguing that PKA-mediated phosphorylation of $Na_V1.1$ and $Na_V1.2$ (TTX-S channels) leads to a reduction in the peak current (see review by [42]) whereas phosphorylation of $Na_V1.8$ by PKA produces an enhancement of the current [12, 13, 15, 25, 27]. It is not at all clear why phosphorylation of presumably the same serine site(s) found in the intracellular loop between transmembrane domains I and II should have the opposite effect on the current amplitude. This remains one of the more interesting questions regarding the modulation of the TTX-R channel(s) in sensory neurons. One possibility is that phosphorylation of this site(s) results in different patterns of channel trafficking. This notion was proposed in recent work [72] wherein $Na_V1.8$ was expressed in Xenopus oocytes; treatment with forskolin enhanced the current conducted by $Na_V1.8$. The forskolin-induced increase was blocked by pretreatment with chloroquine, a presumed inhibitor of vesicular trafficking. These results suggest that the increase in $Na_V1.8$ current may result from PKA-mediated facilitation of the insertion of additional channels into the membrane. Chloroquine by itself had no effect on the current; however, in the absence of additional experiments confirming the role of trafficking, this idea remains speculation.

Furthermore, it is highly likely that the TTX-S sodium channels expressed in sensory neurons contain the serine residues in the intracellular loop between transmembrane domains I and II, but it appears that activators of the cyclic AMP/PKA pathway have little to no effect on the gating of these particular channels (although Scroggs' group has reported an enhancement of TTX-S I_{Na} by PGE_2 in type 3 and type 4 sensory neurons, see [25]). A great deal of information could be gained from site-directed mutagenesis studies of $Na_V1.X$ wherein the serine(s) between domains I and II are sequentially changed to alanines to determine the functional role of each PKA consensus site. This would provide important information regarding the notion whether a single phosphorylation site or perhaps multiple sites (a permissive site or additive effects) play a critical role in modulating the gating of the channel. In this context, the actions of different inflammatory mediators to modulate the conductance of the mutated channels could be assessed as well as the contribution these altered channels make to the firing properties of the action potential. Similarly, if the PKC phosphorylation site(s) is modified, would this then prevent the capacity of PKA to augment the current? Currently, the vast majority of studies have established that modulation of TTX-R I_{Na} is important in the sensitization of the response to a variety of inflammatory mediators, whereas the TTX-S channels look to have only a small (if any) degree of modulation by intracellular signalling cascades. In addition, understanding the modulation of TTX-S and TTX-R sodium channels is complicated by the fact that electrophysiologically separating these "two" currents is difficult because of the overlap in their biophysical properties as well as the differences in expression levels from neuron to neuron (e.g., [73]). Until a selective blocker of TTX-R is developed, examining TTX-S in isolation will be challenging. Such studies could be performed in the $Na_V1.8$-null mouse, however, we are learning that the subtypes of TTX-S sodium channels each have distinctive properties such that measuring the total current will provide few details in how these individual channels contribute to the overall level of excitability in nociceptive sensory neurons.

References

1   Black JA, Liu S, Tanaka M, Cummins TR, Waxman SG (2004) Changes in the expression of tetrodotoxin-sensitive sodium channels within dorsal root ganglia neurons in inflammatory pain. *Pain* 108: 237–247

2   Coggeshall RE, Tate S, Carlton SM (2004) Differential expression of tetrodotoxin-resistant sodium channels $Na_V1.8$ and $Na_V1.9$ in normal and inflamed rats. *Neurosci Lett* 355: 45–48

3   Gold MS (1999) Tetrodotoxin-resistant Na⁺ currents and inflammatory hyperalgesia. *Proc Natl Acad Sci USA* 96: 7645–7649

4   Lai J, Hunter JC, Porreca F (2003) The role of voltage-gated sodium channels in neuro-pathic pain. *Curr Opin Neurobiol* 13: 291–297

5   Lai J, Porreca F, Hunter JC, Gold MS (2004) Voltage-gated sodium channels and hyper-algesia. *Annu Rev Pharmacol Toxicol* 44: 371–397

6   Novakovic SD, Tzoumaka E, McGivern JG, Haraguchi M, Sangameswaran L, Gogas KR, Eglen RM, Hunter JC (1998) Distribution of the tetrodotoxin-resistant sodium channel PN3 in rat sensory neurons in normal and neuropathic conditions. *J Neurosci* 18: 2174–2187

7   Okuse K, Chaplan SR, McMahon SB, Luo ZD, Calcutt NA, Scott BP, Akopian AN, Wood JN (1997) Regulation of expression of the sensory neuron-specific sodium chan-nel SNS in inflammatory and neuropathic pain. *Mol Cell Neurosci* 10: 196–207

8   Porreca F, Lai J, Bian D, Wegert S, Ossipov MH, Eglen RM, Kassotakis L, Novakovic S, Rabert DK, Sangameswaran L et al (1999) A comparison of the potential role of the tetrodotoxin-insensitive sodium channels, PN3/SNS and NaN/SNS2, in rat models of chronic pain. *Proc Natl Acad Sci USA* 96: 7640–7644

9   Waxman SG, Dib-Hajj S, Cummins TR, Black JA (1999) Sodium channels and pain. *Proc Natl Acad Sci USA* 96: 7635–7639

10 Ferreira SH, Nakamura M (1979) I-Prostaglandin hyperalgesia, a cAMP/Ca^{2+} depen-dent process. *Prostaglandins* 18: 179–190

11 Taiwo YO, Bjerknes LK, Goetzl EJ, Levine JD (1989) Mediation of primary afferent peripheral hyperalgesia by the cAMP second messenger system. *Neuroscience* 32: 577–580

12 Gold MS, Reichling DB, Shuster MJ, Levine JD (1996) Hyperalgesic agents increase a tetrodotoxin-resistant Na^+ current in nociceptors. *Proc Natl Acad Sci USA* 93: 1108–1112

13 England S, Bevan S, Docherty RJ (1996) PGE_2 modulates the tetrodotoxin-resistant sodium current in neonatal rat dorsal root ganglion neurones *via* the cyclic AMP-pro-tein kinase A cascade. *J Physiol* 495: 429–440

14 Gold MS, Levine JD, Correa AM (1998) Modulation of TTX-R I_{Na} by PKC and PKA and their role in PGE_2-induced sensitization of rat sensory neurons *in vitro*. *J Neurosci* 18: 10345–10355

15 Fitzgerald EM, Okuse K, Wood JN, Dolphin AC, Moss SJ (1999) cAMP-dependent phosphorylation of the tetrodotoxin-resistant voltage-dependent sodium channel SNS. *J Physiol* 516: 433–446

16 Moskowitz MA (1993) Neurogenic inflammation in the pathophysiology and treatment of migraine. *Neurology* 43 (Suppl 3): S16–20

17 Mossner R, Lesch KP (1998) Role of serotonin in the immune system and in neuroim-mune interactions. *Brain Behav Immun* 12: 249–271

18 Owen DA (1987) Inflammation – histamine and 5-hydroxytryptamine. *Br Med Bull* 43: 256–269

19 Ryan GB, Majno G (1977) Acute inflammation. A review. *Am J Pathol* 86: 183–276

20   Herbert MK, Schmidt RF (1992) Activation of normal and inflamed fine articular afferent units by serotonin. *Pain* 50: 79–88

21   Mense S (1981) Sensitization of group IV muscle receptors to bradykinin by 5-hydroxytryptamine and prostaglandin E_2. *Brain Res* 225: 95–105

22   Treede RD, Meyer RA, Raja SN, Campbell JN (1992) Peripheral and central mechanisms of cutaneous hyperalgesia. *Prog Neurobiol* 38: 397–421

23   Taiwo YO, Heller PH, Levine JD (1992) Mediation of serotonin hyperalgesia by the cAMP second messenger system. *Neuroscience* 48: 479–483

24   Taiwo YO, Levine JD (1992) Serotonin is a directly-acting hyperalgesic agent in the rat. *Neuroscience* 48: 485–490

25   Cardenas CG, Del Mar LP, Cooper BY, Scroggs RS (1997) 5HT4 receptors couple positively to tetrodotoxin-insensitive sodium channels in a subpopulation of capsaicin-sensitive rat sensory neurons. *J Neurosci* 17: 7181–7189

26   Cardenas CG, Del Mar LP, Scroggs RS (1995) Variation in serotonergic inhibition of calcium channel currents in four types of rat sensory neurons differentiated by membrane properties. *J Neurophysiol* 74: 1870–1879

27   Cardenas LM, Cardenas CG, Scroggs RS (2001) 5HT increases excitability of nociceptor-like rat dorsal root ganglion neurons *via* cAMP-coupled TTX-resistant Na(+) channels. *J Neurophysiol* 86: 241–248

28   d'Alcantara P, Cardenas LM, Swillens S, Scroggs RS (2002) Reduced transition between open and inactivated channel states underlies 5HT increased $I_{(Na+)}$ in rat nociceptors. *Biophys J* 83: 5–21

29   Dray A, Bettaney J, Forster P, Perkins MN (1988) Bradykinin-induced stimulation of afferent fibres is mediated through protein kinase C. *Neurosci Lett* 91: 301–307

30   Rang HP, Ritchie JM (1988) Depolarization of nonmyelinated fibers of the rat vagus nerve produced by activation of protein kinase C. *J Neurosci* 8: 2606–2617

31   Burgess GM, Mullaney I, McNeill M, Dunn PM, Rang HP (1989) Second messengers involved in the mechanism of action of bradykinin in sensory neurons in culture. *J Neurosci* 9: 3314–3325

32   Schepelmann K, Messlinger K, Schmidt RF (1993) The effects of phorbol ester on slowly conducting afferents of the cat's knee joint. *Exp Brain Res* 92: 391–398

33   Barber LA, Vasko MR (1996) Activation of protein kinase C augments peptide release from rat sensory neurons. *J Neurochem* 67: 72–80

34   Kamp TJ, Hell JW (2000) Regulation of cardiac L-type calcium channels by protein kinase A and protein kinase C. *Circ Res* 87: 1095–1102

35   MacDonald JF, Kotecha SA, Lu WY, Jackson MF (2001) Convergence of PKC-dependent kinase signal cascades on NMDA receptors. *Curr Drug Targets* 2: 299–312

36   Shearman MS, Sekiguchi K, Nishizuka Y (1989) Modulation of ion channel activity: a key function of the protein kinase C enzyme family. *Pharmacol Rev* 41: 211–237

37   Swope SL, Moss SJ, Raymond LA, Huganir RL (1999) Regulation of ligand-gated ion channels by protein phosphorylation. *Adv Second Messenger Phosphoprotein Res* 33: 49–78

38 Dumont JE, Pecasse F, Maenhaut C (2001) Crosstalk and specificity in signalling. Are we crosstalking ourselves into general confusion? *Cell Signal* 13: 457–463

39 Dzimiri N (2002) Receptor crosstalk. Implications for cardiovascular function, disease and therapy. *Eur J Biochem* 269: 4713–4730

40 Houslay MD (1995) Compartmentalization of cyclic AMP phosphodiesterases, signalling 'crosstalk', desensitization and the phosphorylation of Gi-2 add cell specific personalization to the control of the levels of the second messenger cyclic AMP. *Adv Enzyme Regul* 35: 303–338

41 Matozaki T, Nakanishi H, Takai Y (2000) Small G-protein networks: their crosstalk and signal cascades. *Cell Signal* 12: 515–524

42 Cantrell AR, Catterall WA (2001) Neuromodulation of Na⁺ channels: an unexpected form of cellular plasticity. *Nat Rev Neurosci* 2: 397–407

43 Khasar SG, McCarter G, Levine JD (1999) Epinephrine produces a beta-adrenergic receptor-mediated mechanical hyperalgesia and *in vitro* sensitization of rat nociceptors. *J Neurophysiol* 81: 1104–1112

44 Ferreira SH, Romitelli M, de Nucci G (1989) Endothelin-1 participation in overt and inflammatory pain. *J Cardiovasc Pharmacol* (Suppl) 5: S220–S222

45 Aramori I, Nakanishi S (1992) Coupling of two endothelin receptor subtypes to differing signal transduction in transfected Chinese hamster ovary cells. *J Biol Chem* 267: 12468–12474

46 Douglas SA, Ohlstein EH (1997) Signal transduction mechanisms mediating the vascular actions of endothelin. *J Vasc Res* 34: 152–164

47 Jiang T, Pak E, Zhang HL, Kline RP, Steinberg SF (1996) Endothelin-dependent actions in cultured AT-1 cardiac myocytes. The role of the epsilon isoform of protein kinase C. *Circ Res* 78: 724–736

48 Kasuya Y, Abe Y, Hama H, Sakurai T, Asada S, Masaki T, Goto K (1994) Endothelin-1 activates mitogen-activated protein kinases through two independent signalling pathways in rat astrocytes. *Biochem Biophys Res Commun* 204: 1325–1333

49 Zhou Z, Davar G, Strichartz G (2002) Endothelin-1 (ET-1) selectively enhances the activation gating of slowly inactivating tetrodotoxin-resistant sodium currents in rat sensory neurons: a mechanism for the pain-inducing actions of ET-1. *J Neurosci* 22: 6325–6330

50 Weskamp G, Otten U (1987) An enzyme-linked immunoassay for nerve growth factor (NGF): a tool for studying regulatory mechanisms involved in NGF production in brain and in peripheral tissues. *J Neurochem* 48: 1779–1786

51 Lewin GR, Ritter AM, Mendell LM (1993) Nerve growth factor-induced hyperalgesia in the neonatal and adult rat. *J Neurosci* 13: 2136–2148

52 Rueff A, Mendell LM (1996) Nerve growth factor and NT-5 induce increased thermal sensitivity of cutaneous nociceptors *in vitro*. *J Neurophysiol* 76: 3593–3596

53 Aguayo LG, White G (1992) Effects of nerve growth factor on TTX- and capsaicin-sensitivity in adult rat sensory neurons. *Brain Res* 570: 61–67

54 Garber SS, Hoshi T, Aldrich RW (1989) Regulation of ionic currents in pheochromocytoma cells by nerve growth factor and dexamethasone. *J Neurosci* 9: 3976–3987

55 Mandel G, Cooperman SS, Maue RA, Goodman RH, Brehm P (1988) Selective induction of brain type II Na$^+$ channels by nerve growth factor. *Proc Natl Acad Sci USA* 85: 924–928

56 Omri G, Meiri H (1990) Characterization of sodium currents in mammalian sensory neurons cultured in serum-free defined medium with and without nerve growth factor. *J Membr Biol* 115: 13–29

57 Rudy B, Kirschenbaum B, Greene LA (1982) Nerve growth factor-induced increase in saxitoxin binding to rat PC12 pheochromocytoma cells. *J Neurosci* 2: 1405–1411

58 Zhang YH, Vasko MR, Nicol GD (2002) Ceramide, a putative second messenger for nerve growth factor, modulates the TTX-resistant Na(+) current and delayed rectifier K(+) current in rat sensory neurons. *J Physiol* 544: 385–402

59 Dobrowsky RT, Carter BD (1998) Coupling of the p75 neurotrophin receptor to sphingolipid signaling. *Ann NY Acad Sci* 845: 32–45

60 Dobrowsky RT, Werner MH, Castellino AM, Chao MV, Hannun YA (1994) Activation of the sphingomyelin cycle through the low-affinity neurotrophin receptor. *Science* 265: 1596–1599

61 Liu B, Andrieu-Abadie N, Levade T, Zhang P, Obeid LM, Hannun YA (1998) Glutathione regulation of neutral sphingomyelinase in tumor necrosis factor-alpha-induced cell death. *J Biol Chem* 273: 11313–11320

62 Liu B, Hannun YA (1997) Inhibition of the neutral magnesium-dependent sphingomyelinase by glutathione. *J Biol Chem* 272: 16281–16287

63 Hilborn MD, Vaillancourt RR, Rane SG (1998) Growth factor receptor tyrosine kinases acutely regulate neuronal sodium channels through the src signaling pathway. *J Neurosci* 18: 590–600

64 Levitan IB (1999) It is calmodulin after all! Mediator of the calcium modulation of multiple ion channels. *Neuron* 22: 645–648

65 Saimi Y, Kung C (2002) Calmodulin as an ion channel subunit. *Annu Rev Physiol* 64: 289–311

66 Trudeau MC, Zagotta WN (2003) Calcium/calmodulin modulation of olfactory and rod cyclic nucleotide-gated ion channels. *J Biol Chem* 278: 18705–18708

67 Wen H, Levitan IB (2002) Calmodulin is an auxiliary subunit of KCNQ2/3 potassium channels. *J Neurosci* 22: 7991–8001

68 Zamponi GW (2003) Calmodulin lobotomized: novel insights into calcium regulation of voltage-gated calcium channels. *Neuron* 39: 879–881

69 Herzog RI, Liu C, Waxman SG, Cummins TR (2003) Calmodulin binds to the C terminus of sodium channels Na$_V$1.4 and Na$_V$1.6 and differentially modulates their functional properties. *J Neurosci* 23: 8261–8270

70 Yang D, Gereau RW 4th (2002) Peripheral group II metabotropic glutamate receptors (mGluR2/3) regulate prostaglandin E$_2$-mediated sensitization of capsaicin responses and thermal nociception. *J Neurosci* 22: 6388–6393

71 Yang D, Gereau RW 4th (2004) Group II metabotropic glutamate receptors inhibit cAMP-dependent protein kinase-mediated enhancement of tetrodotoxin-resistant sodium currents in mouse dorsal root ganglion neurons. *Neurosci Lett* 357: 159–162

72 Vijayaragavan K, Boutjdir M, Chahine M (2004) Modulation of $Na_V1.7$ and $Na_V1.8$ peripheral nerve sodium channels by protein kinase A and protein kinase C. *J Neurophysiol* 91: 1556–1569

73 Rizzo MA, Kocsis JD, Waxman SG (1994) Slow sodium conductances of dorsal root ganglion neurons: intraneuronal homogeneity and interneuronal heterogeneity. *J Neurophysiol* 72: 2796–2815

Na$_V$1.8 as a drug target for pain

Lodewijk V. Dekker and David Cronk

Ionix Pharmaceuticals Ltd, 418 Cambridge Science Park, Cambridge CB4 0PA, UK

Introduction

Scope of this article

Adaptive changes in ion channel expression are a normal part of neuronal plasticity and provide a mechanism for nerve cells to respond to changes in their environment [1, 2]. Inappropriate changes in channel expression or lack of control of channel activity can be major contributing factors to development and maintenance of pathological processes. For example, mutations of ion channels (channelopathies) occur in a variety of CNS disease states such as migraine, ataxia and epilepsy [3–5]. Remodeling of channel expression is also a feature of pathological changes associated with aging [6] and chronic pain [7].

The last few years have seen significant advances in the understanding of the molecular changes associated with establishment and maintenance of chronic pain. Particularly fruitful has been the research on members of the voltage-gated sodium channel family, some of which are expressed in sensory neurons associated with nociceptive signalling. In this chapter we focus on the TTX-resistant sodium channel Na$_V$1.8. We discuss its merit as a therapeutic analgesic target and summarise the current status of ion channel screening technology available to identify modulators of this channel.

Pain and nociceptors

Pain is defined as an unpleasant sensory and emotional experience associated with actual or potential tissue damage. Primary nociceptors represent the start of the sensory pain pathway. Normally, these neurons convey noxious sensory information such as high threshold mechanical information or noxious heat or cold signals. Under pathological conditions of inflammation and nerve damage nociceptors become hyperexcitable, leading to spontaneous action potential activity and repeti-

Sodium Channels, Pain, and Analgesia, edited by Kevin Coward and Mark D. Baker

tive firing. These phenomena are thought to contribute to the complaints of spontaneous pain, hyperalgesia (increased response to normally painful stimuli) or allodynia (response to normally innocuous stimuli) that are often associated with the pathology (for reviews see [8–11]).

As specialised sensory neurons able to detect noxious peripheral information, nociceptors are members of a wider population of neurons capable of detecting and relaying all types of peripheral sensory information. Their cell bodies reside in the dorsal root ganglion (DRG) which therefore contains a very heterogeneous functional cell population. Morphologically, DRG neurones can be differentiated by size, having small, medium or large cell bodies both in dissociated culture preparations and in tissue sections. Individual sensory fibres differ in their degree of myelination resulting in neuronal populations with distinct axonal conduction velocities (Aβ, Aδ and C-fibres), providing yet another level of differentiation of sensory neurons. To a large extent, the small neurons with Aδ and C fibre conductivity represent the nociceptor population although there is no absolute association between function and cell size and there is overlap in cell size between neurons with Aβ, Aδ and C characteristics.

Voltage-gated sodium channels

Voltage-gated sodium channels provide an inward current that underlies the upswing of the neuronal action potential. They contain a pore-forming α-subunit and one or two auxiliary β-subunits. Each α-subunit constitutes a large polypeptide of four "ion channel domains" interspersed with cytoplasmic loops. The ion channel domains are highly conserved within the sodium channel family and consist of two pore-forming transmembrane helices and four voltage sensor transmembrane helices [12, 13]. The ion channel domains are organised in a circular fashion around the ion channel pore, which is shaped by the pore-forming helices donated by each ion channel domain.

Nine voltage-gated sodium channel α-subunits have been identified which can be subgrouped on the basis of their sequence homology and on the basis of their sensitivity to tetrodotoxin (TTX), the puffer fish toxin. There are six TTX-sensitive (TTX-s) and three TTX-resistant (TTX-r) α-subunit genes. Distinct patterns of α-subunit expression exist in different cell types/tissues. Each channel subtype shows subtle differences in biophysical properties, including voltage dependence, rate of activation or rate of inactivation suggesting that they make distinct contributions to membrane excitability. Naturally occurring mutants of these channels and genetic deletion studies have revealed the importance of several of the α-subunits for cellular physiology. Thus mutations in $Na_V1.1$ are associated with some forms of epilepsy while mutations in $Na_V1.5$ mutations underlie cardiovascular syndromes such as Brugada syndrome.

In addition to the α-subunit, voltage-gated sodium channels contain one or two β-subunits of which to date four different genes have been cloned. All contain a single membrane spanning domain at the C-terminus and an extracellular IgG-like fold at the N-terminus. β-subunits have been implicated in sodium channel gating, assembly and cell surface expression. They are also implicated in cell–cell and cell–matrix interactions. Genetic deletion of β2 indicates that this subunit regulates sodium channel density and inactivation as well as neuronal excitability [14].

Na$_V$1.8 in nociceptors

Contribution of Na$_V$1.8 to nociceptor signalling

Electrophysiological evidence indicates that DRG neurons display TTX-s as well as TTX-r sodium currents [15–19]. The major TTX-r current has several unique biophysical properties including high thresholds for activation, high thresholds for steady state-inactivation, rapid recovery from inactivation and a slow rate of inactivation (reviewed in [8]). These properties may underlie some of the physiological properties of nociceptors such as their high activation thresholds in response to noxious stimuli, ongoing activity in the presence of sustained depolarisation, sustained spiking upon prolonged depolarisation and broad action potentials [8]. Within the population of neurons with small diameter cell bodies (loosely representing the nociceptor population), distinct TTX-r sodium currents are observed [20, 21]. Molecular cloning from total DRG mRNA revealed two TTX-r sodium channel α-subunits underlying these currents – Na$_V$1.8 (previously known as PN3, SNS or SNS1) and Na$_V$1.9 (NaN or SNS2) [22–24].

When expressed in *Xenopus laevis* oocytes, Na$_V$1.8 generates a slowly-inactivating Na$^+$ current which is resistant to TTX [22, 25], similar to one of the TTX-r sodium currents recorded in small diameter DRG neurons. DRG neurons obtained from mice in which Na$_V$1.8 has been knocked out by genetic deletion lacked this slowly inactivating TTX-r sodium current [26]. Thus Na$_V$1.8 represents the sodium channel α-subunit underlying this TTX-r current in nociceptors. Further studies have provided proof for the notion that Na$_V$1.8 is expressed in nociceptors with C and A fibre conductivity but not in Aα/β low threshold mechanoreceptors [27].

Several lines of evidence indicate that Na$_V$1.8 conveys some unique membrane characteristics and excitability to the cells in which it is expressed. First, C-type DRG neurons can generate TTX-r sodium-dependent action potentials suggesting a significant contribution of TTX-r sodium channels [28]. Second, there is a positive correlation between expression of Na$_V$1.8 (as determined by immunofluorescence microcopy) and action potential rise time and action potential overshoot suggesting that Na$_V$1.8 contributes to these aspects of the action potential [27]. Third, C-type neurons taken from animals in which Na$_V$1.8 has been deleted by genetic manipulation, showed a reduced peak action potential response and a slower rate of depolarisation than wild type neurons [26, 28]. Fourth, functional expression of physiological levels of $Na_V1.8$ in Purkinje cells, which are normally devoid of this sodium channel, alters the action potential activity of these neurons. This is manifested in three ways: increase in the amplitude and duration of action potentials, decrease in the proportion of action potentials that are conglomerate and the number of spikes per conglomerate action potential and production of sustained, pacemaker-like impulse trains in response to depolarisation [29]. Overall $Na_V1.8$ appears to be an important factor in generating and maintaining the action potential in C-type neurons which in turn affects the firing pattern of nociceptors. Thus $Na_V1.8$ plays a specific and essential role in nociceptive signal transduction.

Apart from $Na_V1.8$, nociceptors express a significant number of other voltage-gated sodium channel α-subunits, and the sum total of their activities determines the membrane current carried [10]. As mentioned, $Na_V1.9$ is a second TTX-r sodium channel in small diameter sensory neurons. Furthermore, these cells contain four TTX-sensitive sodium channels, with significant levels of $Na_V1.1$, $Na_V1.6$ and $Na_V1.7$ and detectable levels of $Na_V1.2$ being present. $Na_V1.3$ is present in embryonic DRG neurons and downregulated during development. In contrast to $Na_V1.8$, little is known regarding the individual contributions of these TTX-sensitive sodium channels to nociceptor function. Future genetic targeting of these subunits and the development of subunit specific antagonists will allow more detailed assessment of their function.

Studies on $Na_V1.8$ in pain models

As mentioned above, neuronal hyperexcitability is an important underlying phenomenon in the pathophysiology of pain. In animal models of inflammatory and neuropathic pain, sensory neurons are hyperexcitable compared to normal conditions and often fire spontaneously. $Na_V1.8$ expression and function have been studied extensively in these models. The data indicate that changes occur in $Na_V1.8$ expression under conditions of inflammation and nerve damage. Although the associated hyperexcitability and ectopic discharge of nociceptors may be explained by these changes in $Na_V1.8$, it should be emphasised that $Na_V1.8$ is not the only factor determining excitability and much wider adaptive patterns of gene expression occur under these conditions which overall will determine neuronal activity.

Inflammatory pain models

Inflammatory mediators are capable of sensitising primary afferent neurons. Many molecular mechanisms have been proposed to explain this phenomenon, one being through an action on voltage-dependent ion channels like $Na_V1.8$. Inflammatory mediators are indeed potent modulators of Na$_V$1.8. Modulation of Na$_V$1.8 occurs through changes in the biophysical properties of Na$_V$1.8 and changes in the levels of Na$_V$1.8.

When assayed on DRGs *ex vivo*, prostaglandin E$_2$, serotonin and adenosine affect the magnitude, voltage dependence and rate of activation and inactivation of the TTX-r current in small neurons [30, 31]. These effects are mediated by activation of a number of intracellular signal transduction pathways including the PKA and the PKC pathways [30]. Activation of these pathways results in a hyperpolarising shift in the activation of the current and increases in the rate of activation and inactivation [30]. PKA affects Na$_V$1.8 directly by phosphorylating a set of residues in the intracellular loop between the first and the second conserved subdomain of the channel [32]. Na$_V$1.8 lacking phosphorylatable residues did not respond to elevation of cAMP. At basal levels of cAMP, the threshold of activation of this mutant was higher than that of wildtype Na$_V$1.8, suggesting that under control conditions PKA has a tonic effect on the channel. Thus, phosphorylation of Na$_V$1.8, tonic or induced, provides a mechanism by which extracellular factors (including inflammatory mediators) can regulate the biophysical properties of Na$_V$1.8 and in this way affect nociceptor activation threshold and action potential frequency.

Certain inflammatory conditions affect the levels of Na$_V$1.8 in the sensory neuron. Intraplantar carrageenan, which induces profound hyperalgesia and allodynia, leads to an increase in Na$_V$1.8 mRNA in the small diameter cell bodies after 4 days with concomitant increases in the TTX-r current densities in these cell bodies [33]. By contrast, intraplantar Freunds Adjuvant does not affect mRNA levels in DRG [34]. However, a marked redistribution of Na$_V$1.8 protein occurs to the digital nerves under these conditions such that the proportion of Na$_V$1.8-expressing axons is increased by 2–5 fold [35].

Neuropathic pain models

Table 1 summarises the Na$_V$1.8 expression in various models of neuropathic pain. Ligation of L5 and L6 spinal nerves (spinal nerve ligation model, SNL) leads to a reduction in the levels of Na$_V$1.8 mRNA and protein in the cell bodies of these neurons concomitant with a reduction in the TTX-r current in these neurons [36, 37]. There is a large increase in Na$_V$1.8 immunoreactivity in the sciatic nerve after L5/L6 ligation which is attributed to redistribution of Na$_V$1.8 from the cell bodies of the L4 DRG (which are still intact) to the sciatic nerve [37]. Concomitant with the increased levels of Na$_V$1.8 in the injured axon, C-fibre conduction in these nerves is much more resistant to TTX than in control nerves, implying that after nerve damage more of the afferent activity is dependent on Na$_V$1.8 [37]. Application of Na$_V$1.8 antisense oligonucleotides intrathecally prevents the upregulation of Na$_V$1.8 in the sciatic nerve and eliminates the TTX-r component of the C-fibre conduction. The effect of antisense oligonucleotides suggests that the increased expression of

Table 1 - $Na_V1.8$ regulation in pain models

Injury	Specimen	Assay parameter	Refs
Spinal nerve ligation			
L5 and L6	L5 14 days	$Na_V1.8$ protein completely reduced	[36]
L5 and L6	L4 14 days	$Na_V1.8$ protein not affected	[36]
L5 and L6	L5	TTX-r current reduced	[37]
	L4	TTX-r current not affected	[37]
L5 and L6	Sciatic nerve	Increase in $Na_V1.8$ protein	[37]
L5 and L6	C-fiber 7 days	Increase in TTX-r C-wave (reduced by antisense to $Na_V1.8$)	[37]
L5 and L6	L5	$Na_V1.8$ protein reduced in small neurons, but not large neurons	[61]
Chronic constriction injury and ligation of sciatic nerve			
Nerve ligation	L4 and L5 21 days	$Na_V1.8$ mRNA reduced by 77% or 42% dependent on rat strain	[34]
CCI	L4 14 days	$Na_V1.8$ protein reduced	[36]
CCI	L4 and L5 14 days	$Na_V1.8$ mRNA reduced	[47]
		TTX-r current density reduced	[47]
CCI	L4 and L5 12 days	TTX-r current densities unaffected	[38]
CCI	L4 and L5 14 days	$Na_V1.8$ mRNA unaffected	[38]
		$Na_V1.8$ protein in small neurones, perinuclear, reduced after CCI	[38]
	Sciatic nerve	$Na_V1.8$ protein highly increased in axons	[38]

$Na_V1.8$ in the sciatic axon is not simply a reflection of redistribution of $Na_V1.8$ from the neuronal cell bodies to the axons but that *de novo* translation of $Na_V1.8$ is required to maintain the increased levels in the sciatic axons. It is not clear whether this translation is induced by the nerve injury or is indeed part of a constitutive synthetic process.

Similar phenomena to the ones described above have been observed in the chronic constriction injury (CCI) model of neuropathic pain in which the sciatic nerve is damaged by a ligature. CCI leads to a reduction of $Na_V1.8$ protein in the cell bodies of small DRG neurons of L4 and L5 concomitant with a reduction in TTX-r sodium current density [34, 36]. As in the SNL model, $Na_V1.8$ protein levels are highly increased in the sciatic nerve after CCI, possibly by redistribution of $Na_V1.8$ from the neuronal cell bodies of the uninjured population of neurons that remains after this type of injury [38].

Expression of TTX-r sodium currents/channels has also been studied after axotomy of the sciatic nerve. There is a fundamental difference between the axotomy model and the two models described above in that no intact nerve fibres are present in the sciatic nerve after axotomy. The behavioural response to axotomy is also different from that after SNL or CCI and largely consists of self-mutilation without allodynia and hyperalgesia. The behavioural autotomy resulting from the axotomy may be a reflection of the experience of spontaneous pain. Sciatic axotomy induces downregulation of Na$_V$1.8 mRNA and downregulation of TTX-r current in the small diameter sensory neuron cell bodies [39–42]. Studies in neuromas resulting from nerve injury indicate that Na$_V$1.8 expression in the neuroma is involved in spontaneous nociceptor activity [43]. At least one example exists in which TTX-r sodium channels in DRG cell bodies are upregulated after axotomy [44]. As discussed by Abdulla and Smith [44], sodium channel expression in the cell bodies after axotomy is determined by several factors including the level of inflammation induced by the injury, the loss of retrograde trophic support *in vivo* and conditions during *ex vivo* plating. Differences between studies may be rationalised on the basis of small differences in the balance of these factors, which may favour inflammation-induced Na$_V$1.8 upregulation under certain conditions and Na$_V$1.8 downregulation through loss of trophic support under other conditions.

Changes in the expression of other sodium channels have also been observed in neuropathic pain models. In the spinal nerve ligation model, Na$_V$1.1 and Na$_V$1.2 levels are decreased in dorsal root ganglia (DRG), but Na$_V$1.3 is greatly increased by nerve damage [45, 46]. In the same model, Na$_V$1.8 and Na$_V$1.9 are both down-regulated in injured DRG [34, 36]. Upregulation of Na$_V$1.3 is a general feature of several neuropathic pain models including CCI [47] and streptozotocin-induced diabetic neuropathy [34]. Ectopic activity in sensory neurons after nerve damage is inhibited by tetrodotoxin [48] suggesting a contribution of a TTX-sensitive channel such as Na$_V$1.3, Na$_V$1.6 or Na$_V$1.7 to this phenomenon. A clear demonstration that Na$_V$1.3 has an important role in neuropathic pain comes from recent experiments in a model of spinal cord injury where antisense oligonucleotides specific for Na$_V$1.3 decreased channel expression, reduced hyperexcitability and attenuated mechanical allodynia [45].

Inflamed versus *neuropathic* – *damaged* versus *undamaged*

Although inflammatory and neuropathic pain may be considered separately, in reality there is significant overlap since nerve damage in any of the models described above leads to an inflammatory response which almost by definition will contribute to the pain phenomena. In interpreting the data it is of interest to consider the two broad responses of Na$_V$1.8 after nerve damage, namely downregulation and redistribution. Na$_V$1.8 mRNA and protein is downregulated in the DRGs after ligation and severance of the spinal nerves, after loose ligation of the sciatic nerve and after straight ligation and cut of the sciatic nerve. As discussed, sciatic axotomy results in 100% damage of the nerve fibres whilst SNL (ligation of L5 and L6 spinal nerve) leaves intact sensory fibres from the L4 DRG and CCI leaves intact fibres that have not been severed. The fact that downregulation of $Na_V1.8$ occurs in all three models suggests that it is the damaged neuronal population in SCI and SNL that shows this phenomenon. The most likely explanation is that a loss of retrograde trophic support results in a loss of gene expression. This is essentially confirmed by studies in which trophic factors are restored, resulting in upregulation of $Na_V1.8$ expression [41, 49, 50]. Of interest is the redistribution of $Na_V1.8$ after spinal nerve ligation. Careful studies suggest that it is the uninjured population of fibres from the L4 DRG where the accumulation occurs [37]. Interference studies suggest that this population contributes to nerve conductance. The same phenomena may have occurred after chronic constriction, although formally the redistribution observed here has not been attributed to the population of undamaged fibres [38]. The mechanism by which uninjured fibres can respond to damage of adjacent fibres is most likely through an inflammatory response, induced by the challenge to the nerve. An increasing body of evidence places Wallerian degeneration of severed axons and its associated inflammation at the heart of neuropathic pain [51–53] and it is conceivable that mediators released in this process can provoke $Na_V1.8$ redistribution by acting locally on undamaged fibres [53–55]. Strikingly, only those models in which there is a spared population of intact neurons, display the pain phenomena of hyperalgesia and allodynia. Hence it may be concluded that the redistribution of $Na_V1.8$ over the intact fibres is of particular significance in the initiation and/or maintenance of these particular pain behaviours.

Studies on $Na_V1.8$ in human pain states

Most of the information on adaptive changes to $Na_V1.8$ has been derived from animal models; however recent reports suggest that $Na_V1.8$ is affected in human pain states too. The features of $Na_V1.8$ expression in tissue biopsies from patients parallel those in some of the animal models described above. In brachial plexus injury patients, there was an acute decrease of $Na_V1.8$ (and indeed also of $Na_V1.9$) immunoreactivity in sensory cell bodies of cervical dorsal root ganglia whose central axons had been avulsed from spinal cord, with gradual return of the immunoreactivity to control levels over months [56]. In contrast, there was increased immunoreactivity in some peripheral nerve fibers just proximal to the site of injury in brachial plexus trunks, and in neuromas. These findings suggest that pre-synthesised channel proteins may undergo translocation with accumulation at sites of nerve injury, as in animal models of peripheral axotomy. Nerve terminals in distal limb neuromas and skin from patients with chronic local hyperalgesia and allodynia all showed marked increases of $Na_V1.8$-immunoreactive fibres, similar to what occurs in the SNL and CCI animal models of neuropathic pain. Increased levels of axonal Na$_V$1.8 may therefore be related to the persistent hypersensitive state [56]. Changes in Na$_V$1.8 expression have also been observed in the causalgic finger, which showed a marked increase in the number and intensity of Na$_V$1.8-immunoreactive nerve terminals compared to control tissue. Other proteins, including Nerve Growth Factor, Nerve Growth Factor Receptor (trk A), and Na$_V$1.9 were unaffected in this conditions [57]. Of interest is the fact that expression of sodium channel β1- and β2-subunits also changes in cervical sensory ganglia after avulsion injury, in parallel with the changes described for Na$_V$1.8 [58].

Interference with Na$_V$1.8 function – Na$_V$1.8 deficient mice and antisense treatment

Na$_V$1.8 deficient mice show a pronounced analgesia to noxious mechanical stimulation and mild deficits in noxious thermoreception [26] (Tab. 2). Thermal hyperalgesia associated with inflammation induced by intraplantar carrageenan or systemic NGF treatment is delayed in the Na$_V$1.8 deficient animals suggesting an involvement of Na$_V$1.8 in the development of inflammatory hyperalgesia [26, 59]. However, no effect of the genetic deletion was observed on PGE$_2$-induced thermal hyperalgesia [59]. Similarly, no effect was observed on mechanical allodynia and thermal hyperalgesia after partial ligation of the sciatic nerve [59]. The interpretation of this data is complicated by the fact that TTX-sensitive sodium currents are upregulated in DRG neurons from Na$_V$1.8 deficient animals [26]. The lack of a phenotypic effect in some forms of experimental algesia may be due to this upregulation. The observation that thermal hyperalgesia is completely reversed in Na$_V$1.8 deficient animals in the presence of systemic lidocaine whilst lidocaine has no effect on control animals adds weight to the argument that the contribution of Na$_V$1.8 may to some extent be masked by compensatory increases in other sodium channel subunits and may be more important in the pain models than suggested by the data [26].

An alternative method to establish the relevance of Na$_V$1.8 for pain signalling is the administration of antisense oligonucleotides. Antisense oligonucleotides reduce the TTX-r sodium current in small diameter DRG neurons *ex vivo* without affecting the TTX-sensitive current [60]. They also reduce the levels of Na$_V$1.8 in the cell bodies of DRG neurons after intrathecal administration [11, 61]. Mechanical allodynia resulting from ligation of the L5/L6 spinal nerves is attenuated by intrathecal Na$_V$1.8 antisense oligonucleotides [61]. PGE$_2$-induced and Freund's adjuvant-induced but not carrageenan-induced hyperalgesia is reduced by antisense Na$_V$1.8 confirming that Na$_V$1.8 plays a role in inflammation associated pain [60].

The Na$_V$1.8 knockout and antisense studies present an incomplete and also inconsistent picture. Why for instance is PGE$_2$-induced hyperalgesia reversed by

Table 2 - Behavioural changes in Na$_V$1.8-deficient mice

Nociceptive threshold

Modality	Stimulus	Effect of genetic deletion	Ref
Thermal	Noxious irradiation	Paw flick latency increased	[26]
		Tail flick latency increased	[26]
	Hot plate	No effect on paw flick latency	[26]
Mechanical	Tail pressure	Escape response latency increased	[26]
	Von Frey	No effect	[26]

Inflammatory pain

Challenge	Readout	Effect of genetic deletion	Ref
Intraplantar Carrageenan	Thermal hyperalgesia	Attenuated at short time points, no change at later time points	[26]
Systemic NGF	Thermal hyperalgesia	Attenuated at 6 h, no differences at 24 h	[59]
Intraplantar PGE$_2$	Thermal hyperalgesia	No effect	[59]

Neuropathic pain

Challenge	Readout	Effect of genetic deletion	Ref
Partial ligation of sciatic nerve	Thermal hyperalgesia	No effect	[59]
	Mechanical allodynia	No effect	[59]

Visceral pain

Challenge	Effect of genetic deletion	Ref
Intracolonic capsaicin	Decreased number of behaviours No referred hyperalgesia	[82]
Intracolonic mustard oil	Decreased number of behaviours Weak referred hyperalgesia	[82]
Intraperitoneal cyclophosphamide	No changes in number of behaviours Normal referred hyperalgesia	[82]

antisense oligonucleotides but unaffected by the genetic deletion? Why is neuropathic pain not affected by genetic deletion but inhibited by antisense oligonucleotides? Part of the answer may lie in the fact that, as mentioned, TTX-sensitive currents are upregulated in knockout animals. This affects the interpretation of the data significantly. In the absence of data from the knockout animal, the antisense interference experiments provide the main body of evidence for the role of Na$_V$1.8 in pain pathology. Furthermore, pharmacological studies, using general sodium channel inhibitors indicate a role for sodium channels in pain pathology. Mexiletene, lidocaine and lamotrigine reverse the mechanical allodynia associated with nerve ischaemia [62]. NW-1029 reverses mechanical allodynia induced by chronic inflammation or by chronic constriction of the sciatic nerve [63] and BIII 890 CL and mexiletene reverse mechanical joint hyperalgesia [64]. Inhibition of Na$_V$1.8 may underlie the action of these compounds *in vivo* however these inhibitors lack specificity and effects through other sodium channels may contribute to their antinociceptive action. Further confirmation of the involvement of Na$_V$1.8 in pain pathology requires the availability of suitable pharmacological reagents to interfere specifically and acutely with the channel.

Na$_V$1.8 drug development

Na$_V$1.8 cell lines and expression systems

A prerequisite for drug screening to identify blockers of Na$_V$1.8 are cell lines expressing functional Na$_V$1.8. Attempts have been made to express rat Na$_V$1.8 in COS, CHO or HEK-293 cells with mixed success. After introduction of rat Na$_V$1.8 cDNA, cells expressed either no or low levels of functional current and, if detected, the properties of the expressed channel differed from the endogenous channel in rat DRG neurones [65, 66]. The difficulty of heterologous expression of Na$_V$1.8 may relate to accessory proteins determining the size of the functional population of Na$_V$1.8 sodium channel α-subunits. Generally, sodium channel β-subunits affect the biophysical properties of the α-subunit although β-subunit expression did not affect Na$_V$1.8 functional expression in COS cells (Okuse and Baker, this volume). However, recent reports indicate that it is possible to express rat Na$_V$1.8 in the Ng108 x DRG hybrid cell line ND7.23 and it was argued that this was due to the presence of sodium channel β-subunits in these cells [65, 67]. The annexin II light chain (p11) is a regulatory factor that facilitates the expression of Na$_V$1.8 in sensory neurones [66]. Sensory neurones express high endogenous levels of p11 and antisense downregulation of p11 expression results in a reduction of the Na$_V$1.8 current density in these cells. p11 binds directly to the amino terminus of Na$_V$1.8 and is thought to promote the translocation of Na$_V$1.8 to the plasma membrane, producing functional channels. Importantly, introduction of p11 into CHO cells, which normally have low levels of p11, renders these cells capable of supporting functional Na$_V$1.8, indicating that the presence of p11 is a factor in the cellular context to allow Na$_V$1.8 expression to occur [66]. Thus it may be possible to derive a Na$_V$1.8 expressing cell line in a background of high p11 expression.

Ion channel drug discovery technology

It is widely accepted that conventional electrophysiology represents the "gold standard" for studying ion channel function since it is only under conditions where membrane potential can be controlled that physiologically relevant conditions be created. This applies in particular to voltage-gated ion channels since electrophysiology is the only method where these channels can be gated in a physiological fashion. Conventional electrophysiology is too technically demanding and of insufficient throughput to play a significant role in the early stages of drug discovery. This is changing with the emergence of high-throughput electrophysiology systems (discussed below) but for the most part other functional methods have to be employed to identify compounds that modulate or block ion channel activity. In most cases these methods involve the use of neurotoxins which bind to and modulate the channel. Generally, functional assays which detect changes in ion concentration, either intra- or extracellularly, or membrane potential are favoured over radiolabelled toxin binding assays.

Functional sodium channel assays

In principle, assays for ion channels are straightforward and there are a number of fluorescent and non-fluorescent methods available (Tab. 3) [68] which are amenable to the high-throughput screening environment in 96 and 384-well plate arrays, as well as higher well densities. Non-fluorescent methods directly measure the flux of an ion through the channel of interest, in some cases exploiting the non-selective conductance of ions by the channel under investigation. The ion flux may either be measured by the use of radiotracers (e.g., $^{22}Na^+$ or $[^{14}C]$-guanidinium for sodium channels, $^{86}Rb^+$ for potassium channels) or by atomic absorption spectroscopy (AAS) detection of non-radioactive metal ions. AAS has been widely used for the study of potassium flux through several channels types utilising Rb^+ efflux [69, 70] but methods have recently been developed to utilise Li^+ flux for sodium channels and Ag^+ for investigating chloride channels [71]. The advantage of ion flux measurements is that there is a direct correlation with channel function, although at present, this technology remains restricted to a limited number of channel types.

Development of fluorescence-based assays for sodium ions has been largely hampered by lack of selectivity and sensitivity of sodium ion specific indicators [72]. This has led to the wide-scale adoption of dyes that report changes in membrane potential. In principle these dyes are applicable to all channel types. This is because all ion channels, irrespective of their gating mechanism, share the common features that their activation leads to a change in the ionic balance between the intracellular and extracellular space with a net effect on the membrane potential of the cell. The relative merits of a number of membrane potential dye systems are summarised in Table 4 and have been studied by several groups using several different reader tech-

Table 3 Widely used functional sodium channel high-throughput assay formats

Assay methodology	Advantages	Drawbacks	Approximate throughput[a]
Radiometric ion flux	Direct measure of channel function High sensitivity Good correlation with electrophysiology	Environmental/Safety issues associated with radiochemicals Low temporal resolution High channel expression required	= 1000 wells/h
Non-radiometric flux (Atomic absorption spectroscopy)	Direct measure of channel function High sensitivity Good correlation with electrophysiology	Low temporal resolution	= 3000 wells/h
Fluorescence ion detection	Fast kinetics Sensitivity Cost	Interference from other cell signalling pathways Limited range of selective indicators	10,000 wells/h
Membrane potential sensitive dyes	Widely applicable	Indirect measure of channel function Compound induced artefacts Variable correlation with electrophysiology	= 10,000 wells/h

[a]*Dependent on detection instrument and plate format used*

nologies that are commonplace in the screening environment [73–75]. Use of both 96 and 384-well microplates for these assay formats in screening is commonplace and while the detection systems exist for performing these assays in higher densities, e.g., 1536-well plate assays can be performed on the ImageTrak™ or 1536 and 3456-well plates using the Topology-Compensating Plate Reader (TCPR™), the progression to these higher densities has yet to become routine. As demonstrated by Wolff et al. [75], the fluorescent membrane potential dyes discussed here have similar utility when recording the relatively large changes in membrane potential encountered in depolarisation assays. However, where smaller changes in potential are encountered, such as hyperpolarisation mediated by potassium channel activation, the choice of both dye and detection system may be critical to achieve a workable assay format.

Table 4 - Membrane potential sensitive fluorescent dyes

Dye system	Available readers	Advantages	Disadvantages
Redistribution membrane potential dyes, e.g., $DiBAC_4$	FLIPR™, FlexStation™	Well validated Simple protocol	Slow response time Temperature sensitivity Dye/compound interactions
FLIPR™ membrane potential kit[a]	FLIPR™, FlexStation™, ImageTrak™, Hamamatsu FDSS	Simple protocol Fast kinetics	False hit rate Potential target interference from quenching agent
FRET-based voltage sensor probes dyes[b]	VIPR™, FlexStation™, ImageTrak™, Hamamatsu FDSS	Rapid kinetics Ratiometric Ease of transfer between targets	Assay complexity Cost

[a]*Supplied by Molecular Device Corporation, Sunnyvale CA*
[b]*Supplied by Invitrogen, Carlsbad, CA*
ImageTrak is a registered trademark of PerkinElmer Inc, Boston, MA
VIPR & TCPR are registered trademarks of Aurora Instruments, San Diego, CA
FlexStation & FLIPR are registered trademarks of Molecular Device Corporation, Sunnyvale, CA
Hamatsu FDSS is a registered trademark of Hamamatsu Photonic Systems, Bridgewater, NJ

There are two major issues associated with the use of fluorescence methods for measuring changes on membrane potential. Firstly, the dye-based methods may yield a relatively high false positive and/or negative rate when compared to standard electrophysiology [72]. It is difficult to quantify the extent of this problem as it is likely to be dependent upon the dye system used, the nature of the compounds within the screening collection, and the channel under study. Secondly, for voltage-gated channels the mechanism for activation of the channel in a plate-based assay is artificial, utilising toxins or drugs rather than an electrical impulse. While the first of these can be compensated for by choosing the most appropriate dye system during the assay development phase to show correlation with electrophysiological methods, the second can only be addressed by moving to a plate-based system that enables electrical methods for gating.

High-throughput electrophysiology

The field of high-throughput electrophysiology is probably the area of screening technology that is showing most rapid evolution at present. The impact of applying electrophysiological methods earlier in the drug discovery chain would address the concerns of false hit rate and also provide a physiological mechanism for gating voltage-activated channels [76]. The range of instrumentation, technology employed and cost per data point have been reviewed recently [77]. This article considers the six systems that are either commercially available or in later stage development. There are a number of recent literature articles to support the use of this instrumentation for studying a number of channels in high-throughput fashion [77–79]. While the instrumentation is capable of meeting the throughput demands of the screening environment, if the minimum predicted cost per data point > $1 [77] is realised, the consumables costs of performing such a screen for large compound numbers will be prohibitive for many organisations. It remains more likely that high-throughput electrophysiology systems will permit earlier profiling of hits from screening campaigns, utilising fluorescence methods rather than replacing them, thereby eliminating false positives at an earlier stage.

Perhaps a preferred compromise would be the combination of a fluorescence-based detection system linked with an electrical means for stimulation of the channels. A patent application in 2002 [80] describes a Voltage Ion Plate Reader (VIPR™) with an electrical stimulation head. Data has recently been presented for sodium channels expressed in primary neurons and mammalian cell lines using this system [81]. When comparing compounds known to block sodium channels in a use and frequency-dependent manner the results obtained were comparable to those obtained using standard electrophysiology. It remains to be seen if this electrical stimulation device will be commercialised, but potentially it offers the ion channel screening community the means to activate voltage-gated channels in a physiologically relevant manner while utilising standard laboratory consumables with throughputs comparable to standard plate-based techniques.

Summary

Pain is a complex phenomenon which involves central as well as peripheral neuronal mechanisms. Increased sensitivity and excitability of peripheral nociceptive neurons continues to be regarded as an important contributory event to the onset and/or maintenance of the pain state. As such, molecules that determine nociceptor signalling are prime targets for development of analgesic drugs. Na$_V$1.8 is one of these molecules since (1) its expression is restricted to nociceptors indicating that specific drugs would have a favourable side effect profile, (2) it has an important function in the "normal" nociceptor and is modulated in various ways in pain states suggesting that it may play a role in the pathophysiology of pain, and (3) several interference methods suggest that Na$_V$1.8 contributes to the pathology although each of these have limitations to their interpretability. Technically, Na$_V$1.8 is now a feasible drug target with cell lines being developed that allow high-throughput screening of

Na$_V$1.8 in a physiological context and screening technology being available to measure sodium channel activity in high-throughput functional assays. However, ultimately only the availability of drugs to block Na$_V$1.8 specifically will allow us to assess the merits of this target as an analgesic drug target.

References

1   Birch PJ, Dekker LV, James IF, Southan A, Cronk D (2004) Strategies to identify ion channel modulators: current and novel approaches to target neuropathic pain. *Drug Discov Today* 9: 410–418

2   Waxman SG, Dib-Hajj S, Cummins TR, Black JA (2000) Sodium channels and their genes: dynamic expression in the normal nervous system, dysregulation in disease states (1). *Brain Res* 886: 5–14

3   Jen J (1999) Calcium channelopathies in the central nervous system. *Curr Opin Neurobiol* 9: 274–280

4   Jen JC, Baloh RW (2002) Genetics of episodic ataxia. *Adv Neurol* 89: 459–461

5   Mulley JC, Scheffer IE, Petrou S, Berkovic SF (2003) Channelopathies as a genetic cause of epilepsy. *Curr Opin Neurol* 16: 171–176

6   Toro L, Marijic J, Nishimaru K, Tanaka Y, Song M, Stefani E (2002) Aging, ion channel expression, and vascular function. *Vascul Pharmacol* 38: 73–80

7   Waxman SG, Cummins TR, Black JA, Dib-Hajj S (2002) Diverse functions and dynamic expression of neuronal sodium channels. *Novartis Found Symp* 241: 34–51

8   Gold MS (1999) Tetrodotoxin-resistant Na$^+$ currents and inflammatory hyperalgesia. *Proc Natl Acad Sci USA* 96: 7645–7649

9   Waxman SG, Cummins TR, Dib-Hajj S, Fjell J, Black JA (1999) Sodium channels, excitability of primary sensory neurons, and the molecular basis of pain. *Muscle Nerve* 22: 1177–1187

10   Waxman SG, Dib-Hajj S, Cummins TR, Black JA (1999) Sodium channels and pain. *Proc Natl Acad Sci USA* 96: 7635–7639

11   Porreca F, Lai J, Bian D, Wegert S, Ossipov MH, Eglen RM, Kassotakis L, Novakovic S, Rabert DK, Sangameswaran L et al (1999) A comparison of the potential role of the tetrodotoxin-insensitive sodium channels, PN3/SNS and NaN/SNS2, in rat models of chronic pain. *Proc Natl Acad Sci USA* 96: 7640–7644

12   Catterall WA (1995) Structure and function of voltage-gated ion channels. *Annu Rev Biochem* 64: 493–531

13   Catterall WA (1992) Cellular and molecular biology of voltage-gated sodium channels. *Physiol Rev* 72: S15–S48

14   Chen C, Bharucha V, Chen Y, Westenbroek RE, Brown A, Malhotra JD, Jones D, Avery C, Gillespie PJ, III, Kazen-Gillespie KA et al (2002) Reduced sodium channel density, altered voltage dependence of inactivation, and increased susceptibility to seizures in mice lacking sodium channel b2-subunits. *Proc Natl Acad Sci USA* 99: 17072–17077

15   Petersen M, Pierau FK, Weyrich M (1987) The influence of capsaicin on membrane currents in dorsal root ganglion neurones of guinea-pig and chicken. *Pflugers Arch* 409: 403–410

16   Ogata N, Tatebayashi H (1993) Kinetic analysis of two types of Na$^+$ channels in rat dorsal root ganglia. *J Physiol* 466: 9–37

17   Elliott AA, Elliott JR (1993) Characterization of TTX-sensitive and TTX-resistant sodium currents in small cells from adult rat dorsal root ganglia. *J Physiol* 463: 39–56

18   Roy ML, Narahashi T (1992) Differential properties of tetrodotoxin-sensitive and tetrodotoxin-resistant sodium channels in rat dorsal root ganglion neurons. *J Neurosci* 12: 2104–2111

19   Caffrey JM, Eng DL, Black JA, Waxman SG, Kocsis JD (1992) Three types of sodium channels in adult rat dorsal root ganglion neurons. *Brain Res* 592: 283–297

20   Dib-Hajj SD, Tyrrell L, Cummins TR, Black JA, Wood PM, Waxman SG (1999) Two tetrodotoxin-resistant sodium channels in human dorsal root ganglion neurons. *FEBS Lett* 462: 117–120

21   Cummins TR, Dib-Hajj SD, Black JA, Akopian AN, Wood JN, Waxman SG (1999) A novel persistent tetrodotoxin-resistant sodium current in SNS-null and wild-type small primary sensory neurons. *J Neurosci* 19: RC43

22   Akopian AN, Sivilotti L, Wood JN (1996) A tetrodotoxin-resistant voltage-gated sodium channel expressed by sensory neurons. *Nature* 379: 257–262

23   Sangameswaran L, Delgado SG, Fish LM, Koch BD, Jakeman LB, Stewart GR, Sze P, Hunter JC, Eglen RM, Herman RC (1996) Structure and function of a novel voltage-gated, tetrodotoxin-resistant sodium channel specific to sensory neurons. *J Biol Chem* 271: 5953–5956

24   Dib-Hajj SD, Tyrrell L, Black JA, Waxman SG (1998) NaN, a novel voltage-gated Na channel, is expressed preferentially in peripheral sensory neurons and down-regulated after axotomy. *Proc Natl Acad Sci USA* 95: 8963–8968

25   Sivilotti L, Okuse K, Akopian AN, Moss S, Wood JN (1997) A single serine residue confers tetrodotoxin insensitivity on the rat sensory-neuron-specific sodium channel SNS. *FEBS Lett* 409: 49–52

26   Akopian AN, Souslova V, England S, Okuse K, Ogata N, Ure J, Smith A, Kerr BJ, McMahon SB, Boyce S et al (1999) The tetrodotoxin-resistant sodium channel SNS has a specialized function in pain pathways. *Nat Neurosci* 2: 541–548

27   Djouhri L, Fang X, Okuse K, Wood JN, Berry CM, Lawson SN (2003) The TTX-resistant sodium channel Na$_V$1.8 (SNS/PN3): expression and correlation with membrane properties in rat nociceptive primary afferent neurons. *J Physiol* 550: 739–752

28   Renganathan M, Cummins TR, Waxman SG (2001) Contribution of Na$_V$1.8 sodium channels to action potential electrogenesis in DRG neurons. *J Neurophysiol* 86: 629–640

29   Renganathan M, Gelderblom M, Black JA, Waxman SG (2003) Expression of Na$_V$1.8 sodium channels perturbs the firing patterns of cerebellar Purkinje cells. *Brain Res* 959: 235–242

30   Gold MS, Levine JD, Correa AM (1998) Modulation of TTX-R INa by PKC and PKA and their role in PGE$_2$-induced sensitization of rat sensory neurons *in vitro*. *J Neurosci* 18: 10345–10355

31   Gold MS, Reichling DB, Shuster MJ, Levine JD (1996) Hyperalgesic agents increase a tetrodotoxin-resistant Na$^+$ current in nociceptors. *Proc Natl Acad Sci USA* 93: 1108–1112

32   Fitzgerald EM, Okuse K, Wood JN, Dolphin AC, Moss SJ (1999) cAMP-dependent phosphorylation of the tetrodotoxin-resistant voltage-dependent sodium channel SNS. *J Physiol* 516: 433–446

33   Tanaka M, Cummins TR, Ishikawa K, Dib-Hajj SD, Black JA, Waxman SG (1998) SNS Na$^+$ channel expression increases in dorsal root ganglion neurons in the carrageenan inflammatory pain model. *Neuroreport* 9: 967–972

34   Okuse K, Chaplan SR, McMahon SB, Luo ZD, Calcutt NA, Scott BP, Akopian AN, Wood JN (1997) Regulation of expression of the sensory neuron-specific sodium channel SNS in inflammatory and neuropathic pain. *Mol Cell Neurosci* 10: 196–207

35   Coggeshall RE, Tate S, Carlton SM (2004) Differential expression of tetrodotoxin-resistant sodium channels Na$_V$1.8 and Na$_V$1.9 in normal and inflamed rats. *Neurosci Lett* 355: 45–48

36   Decosterd I, Ji RR, Abdi S, Tate S, Woolf CJ (2002) The pattern of expression of the voltage-gated sodium channels Na$_V$1.8 and Na$_V$1.9 does not change in uninjured primary sensory neurons in experimental neuropathic pain models. *Pain* 96: 269–277

37   Gold MS, Weinreich D, Kim CS, Wang R, Treanor J, Porreca F, Lai J (2003) Redistribution of Na$_V$1.8 in uninjured axons enables neuropathic pain. *J Neurosci* 23: 158–166

38   Novakovic SD, Tzoumaka E, McGivern JG, Haraguchi M, Sangameswaran L, Gogas KR, Eglen RM, Hunter JC (1998) Distribution of the tetrodotoxin-resistant sodium channel PN3 in rat sensory neurons in normal and neuropathic conditions. *J Neurosci* 18: 2174–2187

39   Black JA, Cummins TR, Plumpton C, Chen YH, Hormuzdiar W, Clare JJ, Waxman SG (1999) Upregulation of a silent sodium channel after peripheral, but not central, nerve injury in DRG neurons. *J Neurophysiol* 82: 2776–2785

40   Cummins TR, Waxman SG (1997) Downregulation of tetrodotoxin-resistant sodium currents and upregulation of a rapidly repriming tetrodotoxin-sensitive sodium current in small spinal sensory neurons after nerve injury. *J Neurosci* 17: 3503–3514

41   Dib-Hajj SD, Black JA, Cummins TR, Kenney AM, Kocsis JD, Waxman SG (1998) Rescue of α-SNS sodium channel expression in small dorsal root ganglion neurons after axotomy by nerve growth factor *in vivo*. *J Neurophysiol* 79: 2668–2676

42   Fjell J, Cummins TR, Dib-Hajj SD, Fried K, Black JA, Waxman SG (1999) Differential role of GDNF and NGF in the maintenance of two TTX-resistant sodium channels in adult DRG neurons. *Brain Res* Mol *Brain Res* 67: 267–282

43   Roza C, Laird JM, Souslova V, Wood JN, Cervero F (2003) The tetrodotoxin-resistant Na$^+$ channel Na$_V$1.8 is essential for the expression of spontaneous activity in damaged sensory axons of mice. *J Physiol* 550: 921–926

44 Abdulla FA, Smith PA (2002) Changes in Na$^+$ channel currents of rat dorsal root ganglion neurons following axotomy and axotomy-induced autotomy. *J Neurophysiol* 88: 2518–2529

45 Hains BC, Klein JP, Saab CY, Craner MJ, Black JA, Waxman SG (2003) Upregulation of sodium channel Na$_V$1.3 and functional involvement in neuronal hyperexcitability associated with central neuropathic pain after spinal cord injury. *J Neurosci* 23: 8881–8892

46 Kim CH, Oh Y, Chung JM, Chung K (2001) The changes in expression of three subtypes of TTX sensitive sodium channels in sensory neurons after spinal nerve ligation. *Brain Res* Mol *Brain Res* 95: 153–161

47 Dib-Hajj SD, Fjell J, Cummins TR, Zheng Z, Fried K, LaMotte R, Black JA, Waxman SG (1999) Plasticity of sodium channel expression in DRG neurons in the chronic constriction injury model of neuropathic pain. *Pain* 83: 591–600

48 Omana-Zapata I, Khabbaz MA, Hunter JC, Clarke DE, Bley KR (1997) Tetrodotoxin inhibits neuropathic ectopic activity in neuromas, dorsal root ganglia and dorsal horn neurons. *Pain* 72: 41–49

49 Cummins TR, Black JA, Dib-Hajj SD, Waxman SG (2000) Glial-derived neurotrophic factor upregulates expression of functional SNS and NaN sodium channels and their currents in axotomized dorsal root ganglion neurons. *J Neurosci* 20: 8754–8761

50 Black JA, Langworthy K, Hinson AW, Dib-Hajj SD, Waxman SG (1997) NGF has opposing effects on Na$^+$ channel III and SNS gene expression in spinal sensory neurons. *Neuroreport* 8: 2331–2335

51 Myers RR, Heckman HM, Rodriguez M (1996) Reduced hyperalgesia in nerve-injured WLD mice: relationship to nerve fiber phagocytosis, axonal degeneration, and regeneration in normal mice. *Exp Neurol* 141: 94–101

52 Ramer MS, French GD, Bisby MA (1997) Wallerian degeneration is required for both neuropathic pain and sympathetic sprouting into the DRG. *Pain* 72: 71–78

53 Watkins LR, Maier SF (2002) Beyond neurons: evidence that immune and glial cells contribute to pathological pain states. *Physiol Rev* 82: 981–1011

54 Wu G, Ringkamp M, Hartke TV, Murinson BB, Campbell JN, Griffin JW, Meyer RA (2001) Early onset of spontaneous activity in uninjured C-fiber nociceptors after injury to neighboring nerve fibers. *J Neurosci* 21: RC140

55 Wu G, Ringkamp M, Murinson BB, Pogatzki EM, Hartke TV, Weerahandi HM, Campbell JN, Griffin JW, Meyer RA (2002) Degeneration of myelinated efferent fibers induces spontaneous activity in uninjured C-fiber afferents. *J Neurosci* 22: 7746–7753

56 Coward K, Plumpton C, Facer P, Birch R, Carlstedt T, Tate S, Bountra C, Anand P (2000) Immunolocalization of SNS/PN3 and NaN/SNS2 sodium channels in human pain states. *Pain* 85: 41–50

57 Shembalkar PK, Till S, Boettger MK, Terenghi G, Tate S, Bountra C, Anand P (2002) Increased sodium channel SNS/PN3 immunoreactivity in a causalgic finger. *Eur J Pain* 5: 319–323

58 Coward K, Jowett A, Plumpton C, Powell A, Birch R, Tate S, Bountra C, Anand P

(2001) Sodium channel β1 and β2 subunits parallel SNS/PN3 alpha-subunit changes in injured human sensory neurons. *Neuroreport* 12: 483–488

59 Kerr BJ, Souslova V, McMahon SB, Wood JN (2001) A role for the TTX-resistant sodium channel Na$_V$1.8 in NGF-induced hyperalgesia, but not neuropathic pain. *Neuroreport* 12: 3077–3080

60 Khasar SG, Gold MS, Levine JD (1998) A tetrodotoxin-resistant sodium current mediates inflammatory pain in the rat. *Neurosci Lett* 256: 17–20

61 Lai J, Gold MS, Kim CS, Bian D, Ossipov MH, Hunter JC, Porreca F (2002) Inhibition of neuropathic pain by decreased expression of the tetrodotoxin-resistant sodium channel, Na$_V$1.8. *Pain* 95: 143–152

62 Erichsen HK, Hao JX, Xu XJ, Blackburn-Munro G (2003) A comparison of the antinociceptive effects of voltage-activated Na$^+$ channel blockers in two rat models of neuropathic pain. *Eur J Pharmacol* 458: 275–282

63 Veneroni O, Maj R, Calabresi M, Faravelli L, Fariello RG, Salvati P (2003) Anti-allodynic effect of NW-1029, a novel Na$^+$ channel blocker, in experimental animal models of inflammatory and neuropathic pain. *Pain* 102: 17–25

64 Laird JM, Carter AJ, Grauert M, Cervero F (2001) Analgesic activity of a novel use-dependent sodium channel blocker, crobenetine, in mono-arthritic rats. *Br J Pharmacol* 134: 1742–1748

65 John VH, Main MJ, Powell AJ, Gladwell ZM, Hick C, Sidhu Hs, Clare JJ, Tate S, Trezise DJ (2004) Heterologous expression and functional analysis of rat Na$_V$1.8 (SNS) voltage-gated sodium channels in the dorsal root ganglion neuroblastoma cell line ND7-23. *Neuropharmacology* 46: 425–438

66 Okuse K, Malik-Hall M, Baker MD, Poon WY, Kong H, Chao MV, Wood JN (2002) Annexin II light chain regulates sensory neuron-specific sodium channel expression. *Nature* 417: 653–656

67 Zhou X, Dong XW, Crona J, Maguire M, Priestley T (2003) Vinpocetine is a potent blocker of rat Na$_V$1.8 tetrodotoxin-resistant sodium channels. *J Pharmacol Exp Ther* 306: 498–504

68 Cox B, Denyer JC, Binnie A, Donnelly MC, Evans B, Green DV, Lewis JA, Mander TH, Merritt AT, Valler MJ, Watson SP (2000) Application of high-throughput screening techniques to drug discovery. *Prog Med Chem* 37: 83–133

69 Terstappen GC (1999) Functional analysis of native and recombinant ion channels using a high-capacity nonradioactive rubidium efflux assay. *Anal Biochem* 272: 149–155

70 Scott CW, Wilkins DE, Trivedi S, Crankshaw DJ (2003) A medium-throughput functional assay of KCNQ2 potassium channels using rubidium efflux and atomic absorption spectrometry. *Anal Biochem* 319: 251–257

71 Liang DC (2003) Using flame and graphite furnace atomic absorption spectrometry for analysis of sodium channel activity. *Aurora Biomed Inc.* US PCT application US2003/0100121

72 Xu J, Wang X, Ensign B, Li M, Wu L, Guia A, Xu J (2001) Ion-channel assay technologies: quo vadis? *Drug Discov Today* 6: 1278–1287

73 Gonzalez JE, Tsien RY (1997) Improved indicators of cell membrane potential that use fluorescence resonance energy transfer. *Chem Biol* 4: 269–277

74 Falconer M, Smith F, Surah-Narwal S, Congrave G, Liu Z, Hayter P, Ciaramella G, Keighley W, Haddock P, Waldron G et al (2002) High-throughput screening for ion channel modulators. *J Biomol Screen* 7: 460–465

75 Wolff C, Fuks B, Chatelain P (2003) Comparative study of membrane potential-sensitive fluorescent probes and their use in ion channel screening assays. *J Biomol Screen* 8: 533–543

76 Willumsen NJ, Bech M, Olesen SP, Jensen BS, Korsgaard MP, Christophersen P (2003) High throughput electrophysiology: new perspectives for ion channel drug discovery. *Receptors Channels* 9: 3–12

77 Comley J (2003) Patchers v Screeners: divergent opinion on high-throughput electrophysiology. *Drug Discov World* 4: 47–57

78 Asmild M, Oswald N, Krzywkowski KM, Friis S, Jacobsen RB, Reuter D, Taboryski R, Kutchinsky J, Vestergaard RK, Schroder RL et al (2003) Upscaling and automation of electrophysiology: toward high throughput screening in ion channel drug discovery. *Receptors Channels* 9: 49–58

79 Schroeder K, Neagle B, Trezise DJ, Worley J (2003) Ionworks HT: a new high-throughput electrophysiology measurement platform. *J Biomol Screen* 8: 50–64

80 Maher M, Gonzalez JE (2002) Ion channel assay methods. Aurora Biosciences Corporation PCT international application WO0208748

81 Huang T, Maher MP, Harootunian A, Numann R, Gonzalez JE (2003) Identification of use-dependent blockers of voltage-gated sodium channels with a novel high-throughput optical assay. Society for Neuroscience 33rd Annual Meeting, New Orleans, LA, USA (Abstract 8.10)

82 Laird JM, Souslova V, Wood JN, Cervero F (2002) Deficits in visceral pain and referred hyperalgesia in Na$_V$1.8 (SNS/PN3)-null mice. *J Neurosci* 22: 8352–8356

Role of voltage-gated sodium channels in oral and craniofacial pain

Michael S. Gold

Department of Biomedical Sciences Dental School, Program in Neuroscience, and Department of Anatomy and Neurobiology, Medical School, University of Maryland, Baltimore, MD 21201, USA

Introduction

Voltage-gated sodium channels (VGSCs) are critical for rapid signaling in excitable cells. These channels open in response to membrane depolarization and as their name implies, these open channels enable the influx of sodium. Sodium influx, in turn, causes additional membrane depolarization. The result is a feed-forward reaction responsible for a ~100 mV change in membrane potential that can occur in less then 1 msec. Once opened, the channels quickly transition into an inactive state (a process referred to as inactivation) that enables the membrane potential to be quickly repolarized. The rapid change in membrane potential (depolarization followed by repolarization) is called an action potential and is the fundamental event enabling rapid signaling in the nervous system. It was once thought that VGSCs were merely the driving force behind action potentials, while a myriad of potassium channels were brought into to play to sculpt the various aspects of the action potential (threshold, duration), inter-spike interval, and burst duration. However, it has become clear that the properties of VGSCs are not static and that changes in the properties of these channels contribute to change in excitability.

As detailed in previous chapters, the α-subunit of the channel contains all of the "machinery" necessary for a functional channel including the voltage sensor, ion pore and inactivation gate. The properties of the channel can be influenced by changes in membrane potential, phosphorylation state and β-subunits (i.e., see Chapter by L.V. Dekker and D. Cronk). Ten distinct α-subunits have been identified. Nine of these have been characterized in heterologous expression systems and/or native tissues and appear to possess unique biophysical and/or pharmacological properties [1]. The tenth α-subunit (called Na_X) is a distant relative to the other nine, has yet to be functionally expressed in a heterologous expression system and appears to function as a sodium sensor rather than a voltage-gated channel [2]. Thus, the properties of voltage-gated sodium currents observed at a macroscopic level reflect the unique properties and relative density of the underlying channels. Importantly, it has also become clear that changes in excitability may also reflect changes in the relative density of various VGSCs.

There are four distinct but critical steps that underlie rapid signaling between neurons. The first step involves detection of an incoming signal. This involves stimulus transduction in the peripheral nervous system, whereby a stimulus from the environment (mechanical, thermal or chemical) is converted into an electrical signal called a generator or receptor potential. The analogous event in the central nervous system is called a synaptic potential. The second step involves action potential generation, which results from changes in membrane potential associated with generator or synaptic potentials. The third step involves action potential conduction, where action potentials are conducted along an axon. And the fourth step involves synaptic communication, whereby 1 neuron influences the membrane potential of a second neuron via a chemical synapse or more rarely an electrical synapse. VGSCs are critical for the 2nd and 3rd steps and have been shown to influence the 4th step as well [3]. Given this fact, it follows that changes in the biophysical properties, distribution and/or relative density of VGSCs will have a profound influence on neuronal excitability. For example, a neuron that depends on the relatively high threshold VGSC $Na_V1.8$ for spike initiation would be less excitable than one that depends on a relatively low threshold VGSC such as $Na_V1.7$ or $Na_V1.3$. Thus, all else being equal, a decrease in $Na_V1.8$ and an increase in $Na_V1.3$ at a spike initiation zone would be associated with an increase in excitability.

We presently know the most about VGSCs in peripheral tissue in general and peripheral sensory nerves in particular. This is true for two main reasons. The first reason is convenience, as peripheral tissue is easily accessible, lending itself to experimental manipulation and analysis. The second reason is the appreciation that pain associated with peripheral injury reflects, at least in part, abnormal activity in primary afferent neurons. It is through the effort to understand the basis for injury-induced changes in excitability that researchers have documented the importance of VGSCs [4, 5]. Thus, while there are several lines of evidence suggesting that alterations in VGSCs within the central nervous system may contribute to pain associated with tissue injury [6–8], the focus of this chapter is on the role of these channels in modulating the excitability of primary afferents; specifically primary afferents innervating oral and craniofacial structures.

It has long been appreciated that there is considerable heterogeneity among sensory neurons with respect to electrophysiological, biochemical and morphological properties. However, it is becoming increasingly clear that these neurons are also heterogeneous in their response to injury. Interestingly, this heterogeneity reflects both target of innervation as well as the type of injury. The influence of target of innervation is illustrated by data from models of cystitis [9], ileitis [10, 11], colitis [12] and gastritis [13, 14]. That is, while there are potentially important methodological differences between these models (i.e., method used to induce inflammation, or species studied), all result in a dramatic increase in the excitability of small diameter sensory neurons. However, there are striking differences between them with respect to the underlying mechanism of this increase in excitability. The cystitis model results in a decrease in an inactivating voltage-gated potassium channel (I_A) that reflects a leftward shift in the current availability curve (i.e., fewer channels available to activate at a given membrane potential) [9]. The increase in afferent excitability observed in the ileitis model appears to reflect changes in both voltage-gated Na$^+$ and voltage-gated K$^+$ currents characterized by 1) an increase in TTX-R INa that was associated with no change in the voltage dependence of activation or inactivation, 2) a decrease in I_A that was associated with a leftward shift in the availability curve of the current and 3) and decrease in a sustained K$^+$ current (I_{KD}) [11]. The colitis model results in no apparent changes in potassium currents, a small increase in TTX-S I_{Na} and a significant increase in TTX-R I_{Na} current density that is also associated with a leftward shift in the activation curve for TTX-R I_{Na} (i.e., channels were open at more hyperpolarized potentials) [12]. While the gastritis model results in 1) a significant decrease in I_A that is not associated with a shift in the availability curve [13], 2) no change in I_{KD} [13], 3) a small decrease in TTX-S INa [14], and 4) no change in the density of TTX-R I_{Na}, but rather, a leftward shift in the activation curve of TTX-R I_{Na} [14].

Examples of the heterogeneous response of afferents based the type of injury are found in both visceral and somatic afferents where opposing changes in the expression of VGSCs have been documented. That is, inflammation is generally associated with an increase in the excitability of largely high threshold afferents that reflects, at least in part, an increase in tetrodotoxin (TTX) resistant voltage-gated sodium current (TTX-R I_{Na}) [10, 12, 14–16] (but see [17]). More recently, increases in TTX-sensitive voltage-gated sodium currents (TTX-S I_{Na}) have also been demonstrated in response to inflammation [15]. In contrast, nerve injury is generally associated with an increase in the excitability of largely low threshold afferents [18–20] that appears to reflect a decrease in TTX-R I_{Na} and a change in the properties of TTX-sensitive voltage-gated sodium current (TTX-S I_{Na}); the latter change appears to reflect an increase in the expression of Na$_V$1.3 [21]. Interestingly, nerve injury results in similar changes in sodium currents in high threshold afferents, yet these neurons display little increases in excitability ([22], but see [23]).

As suggested by the brief discussion above and detailed in other chapters in this volume, the majority of what it known about VGSCs and their contribution to pain associated with tissue injury has been obtained through the study of somatic and visceral afferents arising from spinal or dorsal root ganglia (DRG). In the present chapter, I will attempt to summarize what is known about VGSCs in the afferents innervating oral and craniofacial structures as well as their contribution to pain states associated with these structures. While much of this work remains preliminary, observations obtained to date suggest that the role of VGSCs in several of these structures may be distinct from that observed in other parts of the body.

Why focus on oral and craniofacial structures?

If one is concerned about underlying mechanisms of pain, there are three main reasons why oral and craniofacial structures should be considered in isolation relative to other somatic and visceral structures.

First, these structures are innervated by afferents arising from trigeminal ganglia (TG) and there is reason to believe that sensory neurons arising from TG are unique relative to those arising from DRG. Minimally, the two ganglia are composed of different populations of neurons. Both ganglia give rise to high threshold afferents and low threshold thermo- and mechanoreceptors. However, while DRG also contain proprioceptive afferents, the majority of proprioceptive afferents innervating oral and craniofacial structures are located in the mesencephalic nucleus of the 5th cranial nerve rather than the TG [24]. The impact of this difference has yet to be determined, however given the compelling evidence for injury-induced changes in afferent phenotype [25] and cross-excitation within a ganglion [26–28], the impact of these processes may be different in DRG than they are in TG. Another difference between DRG and TG is that there is some somatotopic organization of neuronal cell bodies within TG such that the cell bodies of neurons giving rise to the mandibular, maxillary and ophthalmic branches of the trigeminal nerves are located in distinct regions within the ganglia and there is further sub-organization of neurons within these divisions [24]. That such organization is not observed in DRG provides further support for the suggestion that the impact of cross excitation in DRG and TG will be very different. Furthermore, DRG neurons are derived exclusively from neural crest cells, but TG neurons represent an embryologically heterogeneous population of neurons with some derived from neural crest cells and others from placoidal cells [29]. The implication that sensory neurons arising from neural crest differ from those arising from placoidal cells is suggested by observations highlighting differences between DRG and nodose ganglion (NG, sensory neurons derived exclusively from placoidal cells) which include difference in the co-localization of transmitters and receptors [30–32], differential involvement of receptors underlying the actions of inflammatory mediators [33], differences in the underlying mechanisms of inflammatory mediator-induced sensitization [34–37] as well as differences in the response to tissue injury [13, 14, 38]. Whether TG neurons of different embryological origin have distinct innervation patterns and/or whether differences between these neuronal populations influence the response to injury has yet to be investigated.

Second, oral and craniofacial structures will necessarily require innervation via afferents with unique properties. Sensory transduction for four of the five special senses occurs in the head and all of these specialized structures, the eye, nasal epithelium, auditory canal and cochlea, and gustatory epithelium, possess properties found no where else in the body. For example, the cornea is an avascular structure comprised of a rather homogenous cell population that receives little if any efferent input or typical proprioceptive input [39]. These features alone require that the afferents innervating this structure possess unique properties [40]. In addition to the structures underlying the special senses, specialized structures such as the teeth and the temporomandibular joint (TMJ) are also innervated by TG neurons. Teeth are the only structure in the body that are replaced in their entirety. And while many details of this process have yet to be worked out, what is clear is that innervation of the deciduous teeth is transient and the transition from innervated deciduous teeth to innervated permanent teeth is seamless. Similarly, the TMJ is the only joint in the body that is bi-articulating. This feature raises the possibility that its innervation will be unique, even if sensory innervation is unilateral.

Third, there are a number of pain syndromes that are unique to oral and craniofacial structures. The most common of these is migraine, which is a recurrent, debilitating headache thought to involve sensitization of meningeal and/or craniovascular afferents [41]. Not only does this pain syndrome involve a unique structure, but it appears to have a unique pharmacology; serotonin 1B/D receptor agonists (so called triptans) are the most effective drugs for aborting migraine while they possess minimal efficacy for the treatment of pain arising from other visceral or somatic structures [42]. Interestingly, pain arising from the cornea may be the only other structure for which these drugs appear to have therapeutic efficacy [43] suggesting that corneal and meningeal afferents may possess similar properties. Temporomandibular disorder (TMD) is another relatively common debilitating disorder marked by pain in the TMJ and/or muscles of mastication. Trigeminal neuralgia is a third syndrome characterized by intense unilateral pain that manifests with a number of unique features such as its time course (relatively short bouts of pain), triggers for initiation (light touching of a trigger point is often sufficient to induce an attack), and patterns of spread [44], all three pain syndromes have a higher prevalence in women than in men (but see [45]). However, migraine and TMD pain most often manifest during reproductive years [46, 47], while trigeminal neuralgia is rarely observed in people under 50 [45].

Sodium channels in trigeminal afferents

The first biophysical characterization of voltage-gated sodium currents (I_{Na}) in sensory neurons was performed in 1981 [48]. Results from this study on DRG neurons demonstrated the presence of two distinct classes of I_{Na} that were easily distinguished on the basis of both pharmacology (i.e., TTX sensitivity) and biophysical properties. TTX-S I_{Na} was a low threshold, rapidly activating, rapidly inactivating current with an availability curve that was relatively hyperpolarized (i.e., voltage at which half the channels are available for activation ($V_{1/2}$) was –81 mV). TTX-R I_{Na} was a high threshold, slowly activating, slowly inactivating current with an availability curve that was relatively depolarized (i.e., $V_{1/2}$ was –52 mV). These initial observations obtained from dissociated sensory neurons with a precursor of the patch clamp were later confirmed by a number of investigators using the more high fidelity recording technique of patch clamp [49–51]. Subsequent analysis revealed additional TTX-S and TTX-R currents in DRG neurons including distinct TTX-S I_{Na}s separable according to kinetics of activation, inactivation [52] and recovery from inactivation [21] as well as low threshold TTX-R I_{Na} with both rapid [53, 54] and extremely slow [55] kinetics. In support of this electrophysiological evidence, molecular biological and anatomical evidence indicates that sensory neurons express multiple VGSC α- and β-subunits including: two TTX-R channels ($Na_V1.8$ and $Na_V1.9$) [4], one TTX-insensitive channel ($Na_V1.5$) [56] and multiple TTX-S channels [4], as well as all four β-subunits [57–59].

Results from studies of TG neurons in the late 1980s and early 1990s suggested the presence of both TTX-S and TTX-R I_{Na} in these neurons based on analyses of the action potential waveform [60, 61]. However, it was not until 1999 that the first electrophysiological characterization of I_{Na} in TG neurons was published [62]. Results of this study were consistent with earlier analyses of I_{Na} in spinal ganglia, indicating that there are two major sodium currents in TG neurons: TTX-S I_{Na} and TTX-R I_{Na}. Subsequent molecular biological analysis of TG neurons revealed the presence of $Na_V1.8$ [63], $Na_V1.9$ [64], $Na_V1.5$ [65], $Na_V1.6$ [66] and we have preliminary RT-PCR data suggesting that $Na_V1.1$, 1.2, 1.3, and 1.7 are also expressed in TG. Interestingly, expression of $Na_V1.5$ appears to persist through adulthood in TG neurons [65], while the channel appears to be dramatically downregulated following birth in DRG neurons [56]. The anatomical and/or electrophysiological data collected to date suggests that $Na_V1.8$ and $Na_V1.9$ are localized in the peripheral terminals of unmyelinated axons [67, 68] while $Na_V1.6$ is present in both myelinated and unmyelinated axons [66]. Thus, while the expression pattern of VGSC β-subunits has yet to be investigated in detail, it appears that both the VGSCs expressed and the distribution of expression in TG neurons are largely similar to those observed in DRG neurons.

Putative role in specific pain syndromes

Inflammation

As with the expression and distribution of VGSCs in TGs, evidence to date suggests that there are similarities between DRG and TG neurons with respect to the role of VGSCs in mediating specific pain syndromes. These include both inflammatory and neuropathic pain syndromes. For example, there is compelling evidence from the study of unlabelled DRG neurons *in vitro* as well as specific populations of somatic (i.e., those innervating the hindpaw [69]) and visceral (i.e., those innervating the colon [70]) afferents that $Na_V1.8$ contributes to both the initiation and maintenance of inflammatory hyperalgesia (see [5, 71]): the channel is present in peripheral terminals where it appears to be acutely modulated by inflammatory mediators [69, 70], its modulation is blocked by compounds that attenuate inflammatory hyperalgesia [72] and channel density is increased in the presence of persistent inflammation [12, 14, 15, 73]. $Na_V1.8$ is also present in the peripheral terminals of TG afferents innervating all oral and craniofacial structures studied to date. Specifically, there is preliminary data suggesting this channel is present in the terminals of pulpal afferents [74] and there is compelling evidence suggesting the channel underlies spike initiation in both corneal [68] and meningeal [75] afferents. While it is very likely that this channel is modulated by inflammatory mediators, such modulation has yet to be demonstrated. Furthermore, it has yet to be determined whether persistent inflammation of a specific oral or craniofacial structure results in an increase in the expression of $Na_V1.8$. However, there is evidence that modulation of the channel influences the excitability of TG neurons. That is, nicotine, which has anaesthetic properties, attenuates spike initiation and TTX-R I_{Na} in TG neurons [76]. Furthermore, activation of a cGMP-dependent second messenger cascade attenuates sensitization of meningeal afferents [77] and inhibits TTX-R I_{Na} in TG neurons *in vitro* [78].

It is interesting to note that activation of the cGMP/PKG cascade in TG neurons appears to be primarily inhibitory while it can sensitize or inhibit DRG neurons depending on target of innervation. That is, activation of this second messenger cascade results in hyperalgesia and sensitization of intradermal afferents [79, 80] and inhibition of subcutaneous afferents [80, 81]. Additional experiments are necessary in order to determine whether there are similar subpopulations of TG neurons.

Data from studies of somatic afferents suggest that $Na_V1.3$, $Na_V1.7$ and $Na_V1.9$ may also contribute to inflammatory hyperalgesia. There is an increase in $Na_V1.3$ and $Na_V1.7$ mRNA and a concomitant increase in TTX-S I_{Na} in DRG neurons 4 days following induction of persistent inflammation [15]. An inflammation-induced increase in $Na_V1.7$-like immunoreactivity is also observed in DRG cell bodies within 24 h of a cutaneous injection of complete Freund's Adjuvant (in order to induce persistent inflammation) that persists in a subpopulation of small diameter DRG neurons for more than 2 weeks [82, 83]. More recently, data from a tissue specific knockout of $Na_V1.7$ suggests this channel is critical for both the initiation and maintenance of inflammatory hyperalgesia [84]. An increase in current density associated with an increased $Na_V1.7$ expression may account for its role in the maintenance of inflammatory hyperalgesia. However, neither PKA nor PKC activation appears to contribute to the initiation of inflammatory hyperalgesia as both kinases attenuate $Na_V1.7$-mediated currents in a heterologous expression system [85]. This observation is striking in light of evidence indicating that activation of both pathways are critical for the initiation of inflammatory hyperalgesia [86–88]. Rather, another second messenger pathway such as an extracellular regulated kinase (ERK)-dependent pathway may underlie an inflammation-induced increase in channel density [89], and therefore enable Na$_V$1.7 to contribute to the initiation of inflammatory hyperalgesia. That is, given that NGF is increased in the presence of inflammation [90] and NGF has been shown to activate ERK [91, 92], such a mechanism would also account for the NGF-induced increase in the density of Na$_V$1.7 in neurite endings [93].

There is considerably less evidence in support of a role for Na$_V$1.9 in inflammatory hyperalgesia, and even some evidence to the contrary, given that there is no change in Na$_V$1.9 mRNA observed in the presence of persistent inflammation [15] and a decrease in the proportion of unmyelinated axons with immunohistochemically detectable protein in the presence of inflammation [94]. However, the channel does appear to be modulated by the activation of G-proteins [95] (and fluoride [96]), suggesting that it may contribute to the initiation of inflammatory pain.

Unfortunately, the relative contribution of any of these channels to inflammation-induced changes in the excitability of TG neurons remains largely unknown. That said, there is intriguing preliminary data suggesting that Na$_V$1.7 contributes to pain associated with inflamed teeth and does so in a novel fashion. Pulpal afferents, particularly those terminating at the base of the dentine tubules that appear to mediate the sharp pain associated with dentin hypersensitivity have myelinated axons [97]. Na$_V$1.6 is normally present at nodes of Ranvier in axons arising from both DRG and TG, including those present in myelinated axons innervating teeth. However, in myelinated axons terminating in inflamed teeth, there appears to be a dramatic redistribution of Na$_V$1.7 (to nodes normally occupied by Na$_V$1.6 [98, 99]). The implications of this redistribution have yet to be determined. However, Na$_V$1.7 appears to have several unique biophysical properties, enabling the channel to underlie rapid inward currents in response to slow membrane depolarizations [100]. Consequently, an increase in Na$_V$1.7 at nodes of Ranvier may significantly alter neuronal excitability. It was also suggested that an increase in sodium current density at nodes of Ranvier associated with the redistribution of Na$_V$1.7 may contribute to the relatively common loss of local anaesthetic sensitivity observed in the presence of persistent pulpal inflammation as an increase in current density should increase the safety factor associated with action potential generation at each node of Ranvier [97].

Nerve injury

Nerve injury-induced changes in VGSCs have been well documented in spinal afferents, most notably those comprising the sciatic nerve. While there remains considerable controversy over the relative contribution of specific subpopulations of afferents to the nerve injury-induced changes in nociception (i.e., see [101]), as well as the relative impact of changes in VGSCs on the excitability of afferents [102], particularly, small diameter afferents, there is general agreement on a number of key observations. First, nerve injury results in changes in the expression of multiple VGSC α- and β-subunits, including decreases in the expression of Na$_V$1.8 and

Na$_V$1.9 [21, 103–105], and increases in the expression of Na$_V$1.3 [106, 107] and the β3-subunit [57, 58]. Second, as indicated above, development of the most dramatic change in excitability, i.e., an increase in spontaneous or ectopic activity, observed over the first two weeks following nerve injury, occurs in low threshold afferents [18–20]. And third, at least some of this spontaneous activity reflects the development of membrane oscillations, the upstroke of which appears to depend on activity in TTX-sensitive sodium channels [108, 109] (but see [110]).

In preclinical models of neuropathic pain, injury to oral or craniofacial nerves, at least over the short-term (see below) results in changes similar to those observed following injury to other somatic nerves. These include a decrease in the expression of Na$_V$1.8 [63] and an increase in spontaneous activity [111] that appears to reflect the development of membrane potential oscillations [112]. Similar changes have been observed following transection of both the inferior-alveolar nerve [113] and the infra-orbital nerve [111, 112].

Trigeminal neuralgia (tic douloureux) is a neuropathic pain syndrome associated with oral and craniofacial structures. The involvement of VGSCs in this pain syndrome is suggested by the fact that sodium channel blockers such as carbamazepine have the greatest pharmacological efficacy for the treatment of this pain syndrome [44]. While there is still debate over the nature and cause of this pain syndrome, one prominent hypothesis is that it reflects nerve injury associated with vascular compression of the trigeminal root. The clinical efficacy of microvascular decompression surgery supports this hypothesis. This hypothesis is also supported by morphological changes observed in biopsy specimens obtained from patients suffering from trigeminal neuralgia [44]. Histological analysis reveals areas of demyelination and general disruption of nerve cytoarchitecture. However, as pointed out by Devor and colleagues (2002), demyelination alone is insufficient to account for the pain associated with this syndrome as demyelination is generally associated with action potential conduction failure and therefore anaesthesia, rather than pain. Starting with this premise, Devor and colleagues (2002) have proposed the "ignition" hypothesis to account for a list of 14 distinct features of this pain syndrome. Several of the most notable and unique features include: intense unilateral pain, pain triggered by non-noxious stimuli, pain that outlasts the provoking stimuli, pain that spreads beyond the point of stimulation and the efficacy of some pharmacological interventions (anticonvulsants) but not others (barbiturates). The theory is based on observations made from injury to other peripheral nerves whereby a redistribution of VGSCs in injured axons results in sites of membrane instability reflected in the generation of membrane oscillations [109, 114]. These membrane oscillations result in a situation whereby activity evoked from a trigger zone, or a neighboring axon innervating a trigger zone, sets off a burst of activity. The pain can spread because of "cross talk" within the ganglia [28] where activity in one neuron can evoke activity in another neuron, and/or the development of ephaptic connections at sites of injury [115, 116]. Thus, the paroxysm of pain can outlast the initiating stimulus and spread to affect a larger area through the mechanism of cross-talk. The burst of afferent activity is then terminated by an activity dependent build-up of intracellular calcium that then activates a calcium dependent potassium channel, driving membrane hyperpolarization [117].

While the ignition hypothesis accounts for many of the unique features of trigeminal neuralgia, there are several features that will require further explanation. One feature is that innocuous stimulation is able to ignite a paroxysm of pain. In "normal" tissue, stimulation of low threshold cutaneous afferents will never evoke pain. Furthermore, allodynia, pain evoked by normally innocuous stimuli, appears to reflect sensitization of neurons within the central nervous system (CNS) [118], and this central sensitization appears to require activity in high threshold afferents. Given that the nerve injury associated with trigeminal neuralgia appears to primarily impact heavily myelinated, presumably low threshold, afferents [44], the question arises as to whether high threshold input is needed to establish central sensitization in this condition and if so, where it is arising from. One possibility is that low threshold afferents innervating a trigger point may be electrically coupled to high threshold afferents at either the site of injury and/or within the TG. Data from spinal afferent studies suggests that both forms of coupling are possible [27, 115]. Alternatively, data from animal models of nerve injury suggests that the Wallerian degeneration that occurs following a partial nerve injury is associated with the development of spontaneous activity in high threshold afferents spared by a partial nerve injury [119] and it has been suggested that this activity is sufficient to induce central sensitization and the development of allodynia. The problem with this mechanism, however, is that these pain states are generally associated with ongoing pain and/or bouts of spontaneous pain, whereas trigeminal neuralgia is characterized by brief bouts of pain evoked with low threshold stimulation of a trigger point. There is also the possibility the trigeminal neuralgia reflects, at least in part, pathology within the central nervous system. Another feature of trigeminal neuralgia that is inconsistent with the ignition hypothesis is that pain is rarely evoked during sleep [120]. This observation is also used to support the suggestion that the syndrome reflects a pathology within the CNS. A third feature that is inconsistent with the ignition hypothesis is that the symptoms of trigeminal neuralgia are unique to the cranial nerves [44]. This feature underscores the fact that there is still much to be learned about this syndrome and highlights the fact that the trigeminal system is distinct from spinal systems.

Other differences and/or unique features of the trigeminal system

As suggested above, the response to injuring a nerve innervating an oral or craniofacial structure only shares similarities with spinal nerve injury in the period immediately following the injury. At later time points, the two processes begin to diverge in a number of potentially important ways. First, there is expression of ankyrin G, a multifunctional protein thought to be involved in anchoring VGSCs within the plasma membrane [121, 122]. Data from human peripheral nerve taken from somatic tissue suggests ankyrin G is colocalized with $Na_V1.7$ and $Na_V1.3$ in painful neuromas [123]. Importantly, there is significantly more ankyrin G present in painful neuromas than that observed in normal nerve. In contrast, injury to the inferior alveolar nerve is associated with a decrease in the expression of ankyrin G and this decrease persists for at least 13 weeks [63]. These observations suggest that mechanism controlling the distribution of VGSCs within spinal and trigeminal nerves is distinct. Second, there is the issue of nerve injury-induced changes in VGSCs. While a number of changes in VGSCs have been observed following injury to spinal nerves, $Na_V1.8$ is the only channel studied to date in injured trigeminal nerves [63]. As indicated above, following injury to spinal nerves, there is a dramatic and long lasting decrease in the expression of $Na_V1.8$ across all sizes of injured neurons. In contrast, injury to a trigeminal nerve is associated with a transient decrease in $Na_V1.8$ that is restricted to TG neurons with a small cell body diameter [63]. It has yet to be determined whether there is ultimately an accumulation of $Na_V1.8$ at sites of trigeminal nerve injury as has been demonstrated months after the development of neuromas in spinal nerves [124]. Third, there are differences between spinal nerve injuries and trigeminal nerve injuries that may or may not have to do with changes in VGSCs. These include: 1) the proportion of nerves affected with a greater proportion of injured spinal nerves developing spontaneous activity than is observed following injury to trigeminal nerves [111, 112], although the amount of ectopic activity appears to be dependent on the specific nerve injured [125]; 2) the duration that neurons are affected, as spontaneous activity in spinal nerves is observed over a considerably longer period of time than that observed in injured trigeminal nerve [111]; 3) sympathetic sprouting has been well documented to develop around the somata of injured spinal nerves [126], but not around the somata of injured trigeminal nerves [127, 128].

Preliminary data collected to date suggests there may be at least one additional difference between at least some trigeminal nerves and spinal nerves with respect to the response to persistent inflammation. That is, there is evidence that inflammation of either the masseter muscle or the temporomandibular joint results in an increase in excitability that is associated with either no detectable change in sodium currents [129] or a decrease in sodium currents [130]. These data are in stark contrast to the results of inflammation of the rat hindpaw [15], the ileum [10] and the stomach [14], where there is at least an increase in TTX-R I_{Na}, and often an increase in TTX-S I_{Na} [15]. These differences may simply reflect target of innervation, rather than ganglia of origin, but should be pursued in the future.

Summary and conclusions

Because of their fundamental role in action potential generation, VGSCs are critical for neuronal excitability. Evidence collected over the last 10 years indicates that the biophysical properties, expression pattern and/or distribution of VGSCs are subject to change and that such changes underlie pain associated with injury. The majority of this evidence has come from study of spinal afferents where specific patterns of changes in VGSCs have been well characterized. Even though there is compelling evidence to suggest that there may be differences between oral or craniofacial structures, and somatic and visceral structures with respect to the response to injury, studies of trigeminal nerves have revealed a number of important similarities between the two, including the VGSCs expressed, and their biophysical properties, distribution and functional role in sensory afferents. There are also similarities between spinal and trigeminal nerves with respect to the response to injury. However, there are also important differences, several of which may impact therapeutic interventions employed for the treatment of specific pain syndromes.

Acknowledgments
I would like thank Dr. Danny Weinreich for helpful discussions during the preparation of this manuscript. Some of the work described in this chapter was supported by grants from the National Institutes of Health: P50 AR049555 (NIAMS), P01 NS41384 (NINDS) and RO1 NS044992 (NINDS).

References

1 Yu FH, Catterall WA (2003) Overview of the voltage-gated sodium channel family. *Genome Biol* 4: 207

2 Ogata N, Ohishi Y (2002) Molecular diversity of structure and function of the voltage-gated Na$^+$ channels. *Jpn J Pharmacol* 88: 365–377

3 Gu JG, MacDermott AB (1997) Activation of ATP P2X receptors elicits glutamate release from sensory neuron synapses. *Nature* 389: 749–753

4 Waxman SG, Dib-Hajj S, Cummins TR, Black JA (1999) Sodium channels and pain. *Proc Natl Acad Sci USA* 96: 7635–7639

5 Gold MS (2000) Sodium channels and pain therapy. *Cur Op Anaesthesiol* 13: 565–572

6 Hains BC, Klein JP, Saab CY, Craner MJ, Black JA, Waxman SG (2003) Upregulation of sodium channel Na$_V$1.3 and functional involvement in neuronal hyperexcitability associated with central neuropathic pain after spinal cord injury. *J Neurosci* 23: 8881–8892

7 Hains BC, Saab CY, Klein JP, Craner MJ, Waxman SG (2004) Altered sodium channel

expression in second-order spinal sensory neurons contributes to pain after peripheral nerve injury. *J Neurosci* 24: 4832–4839

8   Max MB, Hagen NA (2000) Do changes in brain sodium channels cause central pain? *Neurology* 54: 544–545

9   Yoshimura N, de Groat WC (1999) Increased excitability of afferent neurons innervating rat urinary bladder after chronic bladder inflammation. *J Neurosci* 19: 4644–4653

10   Moore BA, Stewart TM, Hill C, Vanner SJ (2002) TNBS ileitis evokes hyperexcitability and changes in ionic membrane properties of nociceptive DRG neurons. *Am J Physiol Gastrointest Liver Physiol* 282: G1045–G1051

11   Stewart T, Beyak MJ, Vanner S (2003) Ileitis modulates potassium and sodium currents in guinea pig dorsal root ganglia sensory neurons. *J Physiol* 552: 797–807

12   Beyak MJ, Ramji N, Krol KM, Kawaja MD, Vanner SJ (2004) Two TTX-resistant sodium currents in mouse colonic dorsal root ganglia neurons and their role in colitis induced hyperexcitability. *Am J Physiol Gastrointest Liver Physiol* 287: G845–G855

13   Dang K, Bielefeldt K, Gebhart GF (2004) Gastric ulcers reduce A-type potassium currents in rat gastric sensory ganglion neurons. *Am J Physiol Gastrointest Liver Physiol* 286: G573–G579

14   Bielefeldt K, Ozaki N, Gebhart GF (2002) Experimental ulcers alter voltage-sensitive sodium currents in rat gastric sensory neurons. *Gastroenterology* 122: 394–405

15   Black JA, Liu S, Tanaka M, Cummins TR, Waxman SG (2004) Changes in the expression of tetrodotoxin-sensitive sodium channels within dorsal root ganglia neurons in inflammatory pain. *Pain* 108: 237–247

16   Gold MS, Reichling DB, Shuster MJ, Levine JD (1996) Hyperalgesic agents increase a tetrodotoxin-resistant Na⁺ current in nociceptors. *Proc Natl Acad Sci USA* 93: 1108–1112

17   Yoshimura N, de Groat WC (1997) Plasticity of Na⁺ channels in afferent neurones innervating rat urinary bladder following spinal cord injury. *J Physiol (Lond)* 503: 269–276

18   Liu X, Eschenfelder S, Blenk KH, Janig W, Habler H (2000) Spontaneous activity of axotomized afferent neurons after L5 spinal nerve injury in rats. *Pain* 84: 309–318

19   Liu C, Wall PD, Ben-Dor E, Michaelis M, Amir R, Devor M (2000) Tactile allodynia in the absence of C-fiber activation: altered firing properties of DRG neurons following spinal nerve injury. *Pain* 85: 503–521

20   Liu X, Zhou JL, Chung K, Chung JM (2001) Ion channels associated with the ectopic discharges generated after segmental spinal nerve injury in the rat. *Brain Res* 900: 119–127

21   Cummins TR, Waxman SG (1997) Downregulation of tetrodotoxin-resistant sodium currents and upregulation of a rapidly repriming tetrodotoxin-sensitive sodium current in small spinal sensory neurons after nerve injury. *J Neurosci* 17: 3503–3514

22   Flake NM, Lancaster E, Weinreich D, Gold MS (2004) Absence of an association between axotomy-induced changes in sodium currents and excitability in DRG neurons from the adult rat. *Pain* 109: 471–480

23   Abdulla FA, Smith PA (2001) Axotomy- and autotomy-induced changes in the excitability of rat dorsal root ganglion neurons. *J Neurophysiol* 85: 630–643

24   Capra NF, Dessem D (1992) Central connections of trigeminal primary afferent neurons: topographical and functional considerations. *Crit Rev Oral Biol Med* 4: 1–52

25   Woolf CJ (1996) Phenotypic modification of primary sensory neurons: the role of nerve growth factor in the production of persistent pain. *Philos Trans R Soc Lond B Biol Sci* 351: 441–448

26   Oh EJ, Weinreich D (2002) Chemical communication between vagal afferent somata in nodose Ganglia of the rat and the Guinea pig *in vitro*. *J Neurophysiol* 87: 2801–2807

27   Amir R, Devor M (2000) Functional cross-excitation between afferent A- and C-neurons in dorsal root ganglia. *Neuroscience* 95: 189–195

28   Amir R, Devor M (1996) Chemically mediated cross-excitation in rat dorsal root ganglia. *J Neurosci* 16: 4733–4741

29   Xu H, Federoff H, Maragos J, Parada LF, Kessler JA (1994) Viral transduction of trkA into cultured nodose and spinal motor neurons conveys NGF responsiveness. *Dev Biol* 163: 152–161

30   Zhuo H, Ichikawa H, Helke CJ (1997) Neurochemistry of the nodose ganglion. *Prog Neurobiol* 52: 79–107

31   Zhuo H, Helke CJ (1996) Presence and localization of neurotrophin receptor tyrosine kinase (TrkA, TrkB, TrkC) mRNAs in visceral afferent neurons of the nodose and petrosal ganglia. *Brain Res Mol Brain Res* 38: 63–70

32   Averill S, McMahon SB, Clary DO, Reichardt LF, Priestley JV (1995) Immunocytochemical localization of trkA receptors in chemically identified subgroups of adult rat sensory neurons. *Eur J Neurosci* 7: 1484–1494

33   Moore KA, Taylor GE, Weinreich D (1999) Serotonin unmasks functional NK-2 receptors in vagal sensory neurones of the guinea-pig. *J Physiol (Lond)* 514: 111–124

34   Fowler JC, Wonderlin WF, Weinreich D (1985) Prostaglandins block Ca^{2+}-dependent slow afterhyperpolarization independent of effects on Ca^{2+} influx in visceral afferent neurons. *Brain Res* 345: 345–349

35   Weinreich D (1986) Bradykinin inhibits a slow spike afterhyperpolarization in visceral sensory neurons. *Eur J Pharmacol* 132: 61–63

36   Gold MS, Shuster MJ, Levine JD (1996) Role of a slow Ca^{2+}-dependent slow afterhyperpolarization in prostaglandin E2-induced sensitization of cultured rat sensory neurons. *Neurosci Lett* 205: 161–164

37   Nicol GD, Vasko MR, Evans AR (1997) Prostaglandins suppress an outward potassium current in embryonic rat sensory neurons. *J Neurophysiol* 77: 167–176

38   Bielefeldt K, Ozaki N, Gebhart GF (2002) Mild gastritis alters voltage-sensitive sodium currents in gastric sensory neurons in rats. *Gastroenterology* 122: 752–761

39   Muller LJ, Marfurt CF, Kruse F, Tervo TM (2003) Corneal nerves: structure, contents and function. *Exp Eye Res* 76: 521–542

40   Tanelian DL, Brunson DB (1994) Anatomy and physiology of pain with special reference to ophthalmology. *Invest Ophthalmol Vis Sci* 35: 759–763

41 Moskowitz MA, Bolay H, Dalkara T (2004) Deciphering migraine mechanisms: clues from familial hemiplegic migraine genotypes. *Ann Neurol* 55: 276–280

42 Burstein R (2001) Deconstructing migraine headache into peripheral and central sensitization. *Pain* 89: 107–110

43 May A, Gamulescu MA, Bogdahn U, Lohmann CP (2002) Intractable eye pain: indication for triptans. *Cephalalgia* 22: 195–196

44 Devor M, Amir R, Rappaport ZH (2002) Pathophysiology of trigeminal neuralgia: the ignition hypothesis. *Clin J Pain* 18: 4–13

45 Kitt CA, Gruber K, Davis M, Woolf CJ, Levine JD (2000) Trigeminal neuralgia: opportunities for research and treatment. *Pain* 85: 3–7

46 LeResche L (1997) Epidemiology of temporomandibular disorders: implications for the investigation of etiologic factors. *Crit Rev Oral Biol Med* 8: 291–305

47 MacGregor EA (2004) Oestrogen and attacks of migraine with and without aura. *Lancet Neurol* 3: 354–361

48 Kostyuk PG, Veselovsky NS, Fedulova SA, Tsyndrenko AY (1981) Ionic currents in the somatic membrane of rat dorsal root ganglion neurons – I. Sodium currents. *Neuroscience* 6: 2424–2430

49 Ogata N, Tatebayashi H (1993) Kinetic analysis of two types of Na$^+$ channels in rat dorsal root ganglia. *J Physiol (Lond)* 466: 9–37

50 Elliott AA, Elliott JR (1993) Characterization of TTX-sensitive and TTX-resistant sodium currents in small cells from adult rat dorsal root ganglia. *J Physiol (Lond)* 463: 39–56

51 Roy ML, Narahashi T (1992) Differential properties of tetrodotoxin-sensitive and tetrodotoxin-resistant sodium channels in rat dorsal root ganglion neurons. *J Neurosci* 12: 2104–2111

52 Caffrey JM, Eng DL, Black JA, Waxman SG, Kocsis JD (1992) Three types of sodium channels in adult rat dorsal root ganglion neurons. *Brain Res* 592: 283–297

53 Rush AM, Brau ME, Elliott AA, Elliott JR (1998) Electrophysiological properties of sodium current subtypes in small cells from adult rat dorsal root ganglia. *J Physiol (Lond)* 511: 771–789

54 Scholz A, Appel N, Vogel W (1998) Two types of TTX-resistant and one TTX-sensitive Na$^+$ channel in rat dorsal root ganglion neurons and their blockade by halothane. *Eur J Neurosci (Suppl)* 10: 2547–2556

55 Cummins TR, Dib-Hajj SD, Black JA, Akopian AN, Wood JN, Waxman SG (1999) A novel persistent tetrodotoxin-resistant sodium current In SNS-null and wild-type small primary sensory neurons. *J Neurosci* 19(24): 1–6

56 Renganathan M, Dib-Hajj S, Waxman SG (2002) Na$_{(V)}$1.5 underlies the "third TTX-R sodium current" in rat small DRG neurons. *Brain Res Mol Brain Res* 106: 70–82

57 Shah BS, Stevens EB, Gonzalez MI, Bramwell S, Pinnock RD, Lee K, Dixon AK (2000) beta3, a novel auxiliary subunit for the voltage-gated sodium channel, is expressed preferentially in sensory neurons and is upregulated in the chronic constriction injury model of neuropathic pain. *Eur J Neurosci* 12: 3985–3990

58   Takahashi N, Kikuchi S, Dai Y, Kobayashi K, Fukuoka T, Noguchi K (2003) Expression of auxiliary beta subunits of sodium channels in primary afferent neurons and the effect of nerve injury. *Neuroscience* 121: 441–450

59   Yu FH, Westenbroek RE, Silos-Santiago I, McCormick KA, Lawson D, Ge P, Ferriera H, Lilly J, DiStefano PS, Catterall WA et al (2003) Sodium channel beta4, a new disulfide-linked auxiliary subunit with similarity to beta2. *J Neurosci* 23: 7577–7585

60   Galdzicki Z, Puia G, Sciancalepore M, Moran O (1990) Voltage-dependent calcium currents in trigeminal chick neurons. *Biochem Biophys Res Commun* 167: 1015–1021

61   Hsiung GR, Puil E (1990) Ionic dependencies of tetrodotoxin-resistant action potentials in trigeminal root ganglion neurons. *Neuroscience* 37: 115–125

62   Kim HC, Chung MK (1999) Voltage-dependent sodium and calcium currents in acutely isolated adult rat trigeminal root ganglion neurons. *J Neurophysiol* 81: 1123–1134

63   Bongenhielm U, Nosrat CA, Nosrat I, Eriksson J, Fjell J, Fried K (2000) Expression of sodium channel SNS/PN3 and ankyrin(G) mRNAs in the trigeminal ganglion after inferior alveolar nerve injury in the rat. *Exp Neurol* 164: 384–395

64   Dib-Hajj S, Black JA, Cummins TR, Waxman SG (2002) NaN/Na$_V$1.9: a sodium channel with unique properties. *Trends Neurosci* 25: 253–259

65   Kerr NC, Holmes FE, Wynick D (2004) Novel isoforms of the sodium channels Na$_V$1.8 and Na$_V$1.5 are produced by a conserved mechanism in mouse and rat. *J Biol Chem* 279: 24826–24833

66   Black JA, Renganathan M, Waxman SG (2002) Sodium channel Na$_{(V)}$1.6 is expressed along nonmyelinated axons and it contributes to conduction. *Brain Res Mol Brain Res* 105: 19–28

67   Fjell J, Hjelmstrom P, Hormuzdiar W, Milenkovic M, Aglieco F, Tyrrell L, Dib-Hajj S, Waxman SG, Black JA (2000) Localization of the tetrodotoxin-resistant sodium channel NaN in nociceptors. *Neuroreport* 11: 199–202

68   Brock JA, McLachlan EM, Belmonte C (1998) Tetrodotoxin-resistant impulses in single nociceptor nerve terminals in guinea-pig cornea. *J Physiol (Lond)* 512: 211–217

69   Khasar SG, Gold MS, Levine JD (1998) A tetrodotoxin-resistant sodium current mediates inflammatory pain in the rat. *Neurosci Lett* 256: 17–20

70   Yoshimura N, Seki S, Novakovic SD, Tzoumaka E, Erickson VL, Erickson KA, Chancellor MB, de Groat WC (2001) The involvement of the tetrodotoxin-resistant sodium channel Na$_V$1.8 (pn3/sns) in a rat model of visceral pain. *J Neurosci* 21: 8690–8696

71   Gold MS (1999) Tetrodotoxin-resistant Na$^+$ currents and inflammatory hyperalgesia. *Proc Natl Acad Sci USA* 96: 7645–7649

72   Gold MS, Levine JD (1996) DAMGO inhibits prostaglandin E$_2$-induced potentiation of a TTX-resistant Na$^+$ current in rat sensory neurons *in vitro*. *Neurosci Lett* 212: 83–86

73   Tanaka M, Cummins TR, Ishikawa K, Dib-Hajj SD, Black JA, Waxman SG (1998) SNS Na$^+$ channel expression increases in dorsal root ganglion neurons in the carrageenan inflammatory pain model. *Neuroreport* 9: 967–972

74   Hargreaves KM, Dryden J, Schwarze M, Gracia N, Martin WJ, Flores CM (2001)

Development of a model to evaluate phenotypic plasticity in human nociceptors. *Soc Neurosci Abs* 27: 428

75   Strassman AM, Raymond SA (1999) Electrophysiological evidence for tetrodotoxin-resistant sodium channels in slowly conducting dural sensory fibers. *J Neurophysiol* 81: 413–424

76   Liu L, Zhu W, Zhang ZS, Yang T, Grant A, Oxford G, Simon SA (2004) Nicotine inhibits voltage-dependent sodium channels and sensitizes vanilloid receptors. *J Neurophysiol* 91: 1482–1491

77   Levy D, Strassman AM (2004) Modulation of dural nociceptor mechanosensitivity by the nitric oxide – Cyclic GMP signaling cascade. *J Neurophysiol* 92(2): 766–772

78   Liu L, Yang T, Bruno MJ, Andersen OS, Simon SA (2004) Voltage Gated Ion Channels in Nociceptors: Modulation by cGMP. *J Neurophysiol* 92(4): 2323–2332

79   Aley KO, McCarter G, Levine JD (1998) Nitric oxide signaling in pain and nociceptor sensitization in the rat. *J Neurosci* 18: 7008–7014

80   Vivancos GG, Parada CA, Ferreira SH (2003) Opposite nociceptive effects of the arginine/NO/cGMP pathway stimulation in dermal and subcutaneous tissues. *Br J Pharmacol* 138: 1351–1357

81   Sachs D, Cunha FQ, Ferreira SH (2004) Peripheral analgesic blockade of hypernociception: activation of arginine/NO/cGMP/protein kinase G/ATP-sensitive K^+ channel pathway. *Proc Natl Acad Sci USA* 101: 3680–3685

82   Gould HJ 3rd, England JD, Soignier RD, Nolan P, Minor LD, Liu ZP, Levinson SR, Paul D (2004) Ibuprofen blocks changes in $Na_V1.7$ and 1.8 sodium channels associated with complete Freund's adjuvant-induced inflammation in rat. *J Pain* 5: 270–280

83   Gould HJ 3rd, England JD, Liu ZP, Levinson SR (1998) Rapid sodium channel augmentation in response to inflammation induced by complete Freund's adjuvant. *Brain Res* 802: 69–74

84   Nassar MA, Stirling LC, Forlani G, Baker MD, Matthews EA, Dickenson AH, Wood JN (2004) Nociceptor-specific gene deletion reveals a major role for $Na_V1.7$ (PN1) in acute and inflammatory pain. *Proc Natl Acad Sci USA* 101: 12706–12711

85   Vijayaragavan K, Boutjdir M, Chahine M (2004) Modulation of $Na_V1.7$ and $Na_V1.8$ peripheral nerve sodium channels by protein kinase A and protein kinase C. *J Neurophysiol* 91: 1556–1569

86   Gold MS, Levine JD, Correa AM (1998) Modulation of TTX-R I_{Na} by PKC and PKA and their role in PGE_2-induced sensitization of rat sensory neurons *in vitro*. *J Neurosci* 18: 10345–10355

87   Khasar SG, McCarter G, Levine JD (1999) Epinephrine produces a beta-adrenergic receptor-mediated mechanical hyperalgesia and *in vitro* sensitization of rat nociceptors. *J Neurophysiol* 81: 1104–1112

88   Taiwo YO, Bjerknes LK, Goetzl EJ, Levine JD (1989) Mediation of primary afferent peripheral hyperalgesia by the cAMP second messenger system. *Neuroscience* 32: 577–580

89   Wada A, Yanagita T, Yokoo H, Kobayashi H (2004) Regulation of cell surface expres-

sion of voltage-dependent $Na_V1.7$ sodium channels: mRNA stability and posttranscriptional control in adrenal chromaffin cells. *Front Biosci* 9: 1954–1966

90 Shu XQ, Mendell LM (1999) Neurotrophins and hyperalgesia. *Proc Natl Acad Sci USA* 96: 7693–7696

91 Obata K, Yamanaka H, Dai Y, Tachibana T, Fukuoka T, Tokunaga A, Yoshikawa H, Noguchi K (2003) Differential activation of extracellular signal-regulated protein kinase in primary afferent neurons regulates brain-derived neurotrophic factor expression after peripheral inflammation and nerve injury. *J Neurosci* 23: 4117–4126

92 Delcroix JD, Valletta JS, Wu C, Hunt SJ, Kowal AS, Mobley WC (2003) NGF signaling in sensory neurons: evidence that early endosomes carry NGF retrograde signals. *Neuron* 39: 69–84

93 Toledo-Aral JJ, Moss BL, He ZJ, Koszowski AG, Whisenand T, Levinson SR, Wolf JJ, Silos-Santiago I, Halegoua S, Mandel G (1997) Identification of PN1, a predominant voltage-dependent sodium channel expressed principally in peripheral neurons. *Proc Natl Acad Sci USA* 94: 1527–1532

94 Coggeshall RE, Tate S, Carlton SM (2004) Differential expression of tetrodotoxin-resistant sodium channels $Na_V1.8$ and $Na_V1.9$ in normal and inflamed rats. *Neurosci Lett* 355: 45–48

95 Baker MD, Chandra SY, Ding Y, Waxman SG, Wood JN (2003) GTP-induced tetrodotoxin-resistant Na^+ current regulates excitability in mouse and rat small diameter sensory neurones. *J Physiol* 548: 373–382

96 Coste B, Osorio N, Padilla F, Crest M, Delmas P (2004) Gating and modulation of presumptive $Na_V1.9$ channels in enteric and spinal sensory neurons. *Mol Cell Neurosci* 26: 123–134

97 Sorensen HJ, Beeler JJ, Johnson LR, Kleier DJ, Levinson SR, Henry MJ (2003) $Na_V1.7$/Pn1 sodium channel upregulation and accumulation at demyelinated sites in painful human tooth pulp. *Soc Neurosci Abs* 175.13

98 Krzemien DM, Schaller KL, Levinson SR, Caldwell JH (2000) Immunolocalization of sodium channel isoform NaCh6 in the nervous system. *J Comp Neurol* 420: 70–83

99 Tzoumaka E, Tischler AC, Sangameswaran L, Eglen RM, Hunter JC, Novakovic SD (2000) Differential distribution of the tetrodotoxin-sensitive rPN4/NaCh6/Scn8a sodium channel in the nervous system. *J Neurosci Res* 60: 37–44

100 Herzog RI, Cummins TR, Ghassemi F, Dib-Hajj SD, Waxman SG (2003) Distinct repriming and closed-state inactivation kinetics of $Na_V1.6$ and $Na_V1.7$ sodium channels in mouse spinal sensory neurons. *J Physiol* 551: 741–750

101 Gold MS (2000) Spinal nerve ligation: what to blame for the pain and why. *Pain* 84: 117–120

102 Flake NM, Lancaster E, Weinreich D, Gold MS (2004) Absence of an association between axotomy-induced changes in sodium currents and excitability in DRG neurons from the adult rat. *Pain* 109: 471–480

103 Dib-Hajj S, Black JA, Felts P, Waxman SG (1996) Down-regulation of transcripts for Na

channel alpha-SNS in spinal sensory neurons following axotomy. *Proc Natl Acad Sci USA* 93: 14950–14954

104 Decosterd I, Ji RR, Abdi S, Tate S, Woolf CJ (2002) The pattern of expression of the voltage-gated sodium channels Na$_{(V)}$1.8 and Na$_{(V)}$1.9 does not change in uninjured primary sensory neurons in experimental neuropathic pain models. *Pain* 96: 269–277

105 Gold MS, Weinreich D, Kim CS, Wang R, Treanor J, Porreca F, Lai J (2003) Redistribution of Na$_{(V)}$1.8 in uninjured axons enables neuropathic pain. *J Neurosci* 23: 158–166

106 Waxman SG, Kocsis JD, Black JA (1994) Type III sodium channel mRNA is expressed in embryonic but not adult spinal sensory neurons, and is reexpressed following axotomy. *J Neurophysiol* 72: 466–470

107 Black JA, Cummins TR, Plumpton C, Chen YH, Hormuzdiar W, Clare JJ, Waxman SG (1999) Upregulation of a silent sodium channel after peripheral, but not central, nerve injury in DRG neurons. *J Neurophysiol* 82: 2776–2785

108 Amir R, Michaelis M, Devor M (1999) Membrane potential oscillations in dorsal root ganglion neurons: role in normal electrogenesis and neuropathic pain. *J Neurosci* 19: 8589–8596

109 Amir R, Liu CN, Kocsis JD, Devor M (2002) Oscillatory mechanism in primary sensory neurones. *Brain* 125: 421–435

110 Liu CN, Devor M, Waxman SG, Kocsis JD (2002) Subthreshold oscillations induced by spinal nerve injury in dissociated muscle and cutaneous afferents of mouse DRG. *J Neurophysiol* 87: 2009–2017

111 Tal M, Devor M (1992) Ectopic discharge in injured nerves: comparison of trigeminal and somatic afferents. *Brain Res* 579: 148–151

112 Cherkas PS, Huang TY, Pannicke T, Tal M, Reichenbach A, Hanani M (2004) The effects of axotomy on neurons and satellite glial cells in mouse trigeminal ganglion. *Pain* 110: 290–298

113 Bongenhielm U, Yates JM, Fried K, Robinson PP (1998) Sympathectomy does not affect the early ectopic discharge from myelinated fibres in ferret inferior alveolar nerve neuromas. *Neurosci Lett* 245: 89–92

114 Amir R, Michaelis M, Devor M (2002) Burst discharge in primary sensory neurons: triggered by subthreshold oscillations, maintained by depolarizing afterpotentials. *J Neurosci* 22: 1187–1198

115 Amir R, Devor M (1992) Axonal cross-excitation in nerve-end neuromas: comparison of A- and C-fibers. *J Neurophysiol* 68: 1160–1166

116 Rasminsky M (1980) Ephaptic transmission between single nerve fibres in the spinal nerve roots of dystrophic mice. *J Physiol (Lond)* 305: 151–169

117 Amir R, Devor M (1997) Spike-evoked suppression and burst patterning in dorsal root ganglion neurons of the rat. *J Physiol (Lond)* 501: 183–196

118 Meyer RA, Campbell JN, Raja SN (1994) Peripheral neural mechanisms of nociception. In: Wall PD, Melzack R (eds): *Textbook of Pain*. Churchill Livingstone, New York, 13–56

119 Wu G, Ringkamp M, Murinson BB, Pogatzki EM, Hartke TV, Weerahandi HM, Camp-

bell JN, Griffin JW, Meyer RA (2002) Degeneration of myelinated efferent fibers induces spontaneous activity in uninjured C-fiber afferents. *J Neurosci* 22: 7746–7753

120 Canavero S, Bonicalzi V, Pagni CA (1995) The riddle of trigeminal neuralgia. *Pain* 60: 229–231

121 Rasband MN, Peles E, Trimmer JS, Levinson SR, Lux SE, Shrager P (1999) Dependence of nodal sodium channel clustering on paranodal axoglial contact in the developing CNS. *J Neurosci* 19: 7516–7528

122 Kordeli E, Lambert S, Bennett V (1995) AnkyrinG. A new ankyrin gene with neural-specific isoforms localized at the axonal initial segment and node of Ranvier. *J Biol Chem* 270: 2352–2359

123 Kretschmer T, England JD, Happel LT, Liu ZP, Thouron CL, Nguyen DH, Beuerman RW, Kline DG (2002) Ankyrin G and voltage gated sodium channels colocalize in human neuroma – key proteins of membrane remodeling after axonal injury. *Neurosci Lett* 323: 151–155

124 Coward K, Plumpton C, Facer P, Birch R, Carlstedt T, Tate S, Bountra C, Anand P (2000) Immunolocalization of SNS/PN3 and NaN/SNS2 sodium channels in human pain states. *Pain* 85: 41–50

125 Michaelis M, Liu X, Janig W (2000) Axotomized and intact muscle afferents but no skin afferents develop ongoing discharges of dorsal root ganglion origin after peripheral nerve lesion. *J Neurosci* 20: 2742–2748

126 McLachlan EM, Jang W, Devor M, Michaelis M (1993) Peripheral nerve injury triggers noradrenergic sprouting within dorsal root ganglia. *Nature* 363: 543–546

127 Bongenhielm U, Boissonade FM, Westermark A, Robinson PP, Fried K (1999) Sympathetic nerve sprouting fails to occur in the trigeminal ganglion after peripheral nerve injury in the rat. *Pain* 82: 283–288

128 Benoliel R, Eliav E, Tal M (2001) No sympathetic nerve sprouting in rat trigeminal ganglion following painful and non-painful infraorbital nerve neuropathy. *Neurosci Lett* 297: 151–154

129 Harriott A, Kirifides M, Dessem D, Gold MS (2004) Inflammation-induced increase in the excitability of masseter muscle afferents. *J Pain* 5: 13

130 Flake N, Gold MS (2003) Sex differences in voltage-gated sodium currents in sensory neurons innervating the temporomandibular joint. *J Dent Res* 82 (Special Issue A): 1181

Sodium channel gating and drug blockade

Andreas Scholz

Physiologisches Institut, Universität Giessen, Aulweg 129, 35392 Giessen, Germany

Introduction

Despite widespread use of local anaesthetics for well over a century, the molecular mechanisms by which they alter specific peripheral nervous system functions have remained unclear for a long time. Nowadays the sodium (Na^+) channel protein is identified as a target for specific, clinically important, local anaesthetic effects in neurons. The recent findings which have grown from the description of the amino acid sequence of the Na^+ channel protein [1] described the structure of the Na^+ channel with regards to its functions. Molecular mechanisms of local anaesthetic block have been suggested and identified based on this structural information and by earlier observations of the mode of operation of local anaesthetic. However it should be kept in mind that Na^+ channels are not the only targets of local anaesthetics at clinically relevant concentrations; potassium (K^+) and calcium (Ca^{2+}) channels are also affected, which might explain some of the side effects of local anaesthetics. Recent findings indicate that local anaesthetics also act on intracellular mechanisms, which raises the question of whether these might explain toxicity.

Historical view

The first local anaesthetic was cocaine, used by Carl Koller and Sigmund Freud who noticed a numbing effect after applying cocaine on the tongue. Koller, who was intent on finding a drug to anaesthetise the cornea for his ophthalmologic work in Vienna, knew that Freud had relieved pain with cocaine [2]. They demonstrated its local anaesthetic effects with self-experimentation and showed that within minutes they could not feel sensation even when a needle was applied to their cornea. Thereafter, they reported that they were able to painlessly enucleate a dog's eye. Leonard Corning, a neurologist in New York City, injected a cocaine solution (2%) between the spinous process in a young dog which resulted in insensibility within 5 min [3]; this was later applied to patients, with the drug presumably acting in the epidural space. Spinal anaesthesia was introduced later after the demonstration of lumbar puncture by Quincke [4, 5]. Cocaine was widely used despite its disadvantages of high toxicity, short duration of anaesthesia, difficulties involved in sterilising the solution and its costs (not to mention addiction). Alfred Einhorn synthesised procaine after investigating degradation products of cocaine since he reported that "...the anesthetic capability of cocaine is a function of its acid group called by Paul Ehrlich the 'anesthesiophoric' group – the most potent being the benzoyl group." [6]. This structural starting point was the basis of the majority of clinically used local anaesthetics consisting of the benzene ring linked *via* an amide or ester to an amine group, and their names still end in "caine".

Differential and use-dependent block by local anaesthetics

Using local anaesthetics, the concept of "differential nerve block" was noted with sensitivity for block of nerve fibres increasing from sharp pain, cold, warmth and contact or touch, to finally motor fibres [5]. This was subjected to quantitative neurophysiological analysis by Gasser and Erlanger in 1929 when they reported on the differential susceptibility of compound action potentials in nerves to pressure and cocaine-containing solutions based on fibre size and conduction velocity, from Aα (fastest) to C-fibres (slowest) [7]. They suspected that diameter might be the main parameter accounting for differential nerve block, because they felt the process responsible for impulse propagation was essentially similar in all fibres. With cocaine they observed that small fibres (slowly conducting) tended to be blocked before large ones, but in all cases a varying proportion of large fibres were blocked well before the compound action potential for small fibres had disappeared. They concluded that factors other than fibres size must be operating to produce the differential blockade: "Such a simple mechanism as has just been described should cause the fibres to be blocked systematically on a size basis; and since this does not rigidly hold the problem can be considered to be only partly solved. Some other as yet undetermined factor must be operating" [7]. The possibility still exists that the differential blockade is due to the fact that the extent to which the Na$^+$ current must be diminished before block occurs varies between different fibres (i.e., they have different safety factors for conduction). Later experiments with compound action potentials of peripheral nerves of various species revealed that a portion of C-fibres are blocked at the time when A-fibre blockade is observed. However, the concentration necessary for half-maximal block of the C-fibres compound action potential is two to four times higher than that for A-fibres depending on type of local anaesthetic used [8–10]. Taken together from experiments with various local anaesthetics it could be postulated that local anaesthetics with an ester structure have an inherently higher efficacy than those with an amide structure.

Structure

After isolation and purification, the Na^+ channel protein was described as a single polypeptide chain with a relative molecular mass of ~260,000 [11], and was referred to as the α-subunit [12]. While there are 11 genes coding for different α-subunits, only nine are known to be expressed in nature in rats and humans [13, 14]. The polypeptide chain of ~1,950 amino acid residues crosses the cell membrane 24 times. The channel consists of four domains (D I–D IV, Fig. 1a) [15], each comprising six transmembrane helical segments, which are numbered S1–S6. The link between the S5 and S6 in each domain is of particular interest because these "pore loops" form the outer pore and contain the amino acid sequence DEKA (occurring once in each domain). This sequence forms the selectivity filter, permitting primarily Na^+ ions to pass (Fig. 1a). The region of the outer pore mouth is also involved in the binding of toxins like tetrodotoxin, batrachotoxin and conotoxins [16–18]; batrachotoxin, particularly, although it binds at the outer pore mouth, seems to influence the binding of local anaesthetics at the inner pore regions [19–21].

Three auxiliary subunits (β1–3) influence the activation and inactivation parameters of an expressed a-subunit or the level of channel protein expression [22–24]. The structural part of the channel protein (α-subunit) which underlies "fast" inactivation is the intracellular link between D III and D IV (Fig. 1b). Another type of inactivation, the slow "C-type" [25], appears to involve the pore loops mentioned above.

Function

A simple model of Na^+ channel function contains three steps [26, 27]. Firstly a closed state, at potentials negative to –70 mV. The pore in the channel is occluded so that no Na^+ ion can pass from one side to the other. Based upon experiments with K^+ channels utilising patch-clamp recordings, voltage-clamp fluorimetry and spin labelling, it is proposed that the outer pore interacts directly with the S4 voltage sensor to keep the pore occluded or closed [28, 29]. Since, the structure-function motifs in the K^+ channel parallel those in the Na^+ channel, a similar gating mechanism may be operating.

The open state of the channel is initiated by a depolarisation of the transmembrane potential to the threshold potential (usually more positive than –60 to –40 mV, depending on neurone type). The open channel permits Na^+ ions to pass through the pore causing an inward current which propagates depolarisation; this underlies the upstroke of the action potential in most excitable cells (for more details see chapters by J.A. Black et al. and J.A. Brock). During channel opening, the S4 segment twists back driven by both the change in membrane potential difference and by the intrinsic charge changes, to allow the outer pore mouth to expand – resulting in a 20° twist of the α-helix [13].

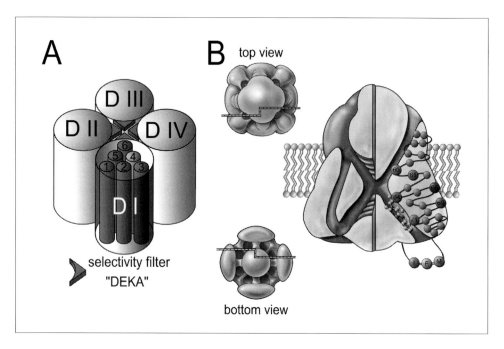

Figure 1

A: schematic diagram of the structure of an α-subunit of a Na⁺ channel. Each of the four domains (D I–D IV) consists of six segments which span the membrane. Part of the "pore" loops, the amino acid link between S5 and S6 segment, is specially highlighted because four amino acids "DEKA" form in the outer pore mouth and act as the selectivity filter (for more details, see the Function and structure of Na⁺ channels *section earlier). B: modified sketch of the 3D view of a Na⁺ channel in cross section. From data of cryo-electron microscopy and single particle analysis, it was possible to derive a 3D model of the Na⁺ channel [74]; top and bottom view show a cross-section of the large sketch. The S4 segment is thought to be on the left side of the cross-section (the function of the cavity marked in red is unclear up to now). The S6 segment is thought to be on the right side of the cross-section and is shown here overlaid with the amino-acid sequence of rat brain Na⁺ channel (Na$_V$1.2). Those residues which are important for the affinity of local anaesthetics are coloured and numbered (60 for amino acid 1760 etc.). (B adapted from [75, 33] with permission from* Nature *and* American Association for Advancement of Science.*]*

After opening (notably during prolonged depolarisation) the channel enters into an inactivated state. During depolarisation the macroscopic Na⁺ current reaches a peak in amplitude and the current amplitude decreases with time, often mono-exponentially (Fig. 2). This self-decreasing inward current is one of the reasons why repolarisation follows an action potential (together with the outward currents

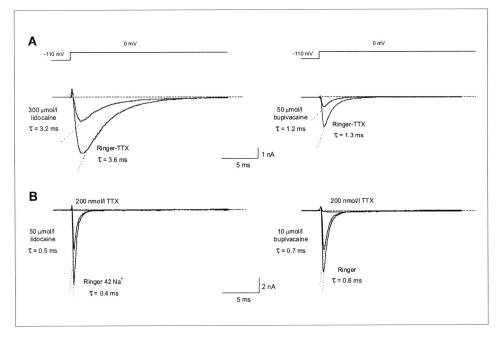

Figure 2
TTX-sensitive and TTX-resistant Na⁺ currents in rat DRG neurons blocked by lidocaine and bupivacaine. A, Extrapolation of the decaying parts of TTX-resistant Na⁺ currents (dotted lines) revealed a time constant of 3.6 ms in Ringer-TTX solution and 3.2 ms in the presence of 300 µmol/l lidocaine. B, The TTX-sensitive Na⁺ currents decayed faster with a time constant of 0.4 ms in Ringer solution and 0.5 ms in the presence of 50 µmol/l lidocaine. Less than 2% of the current remained in the presence of 200 nmol/l TTX (adapted from [41] with permission of J Neurophysiol).

through voltage gated K⁺ channels). Internally applied pronase prevents this "fast" inactivation process, therefore it is concluded that the inactivation gate is positioned on the inner side of the channel protein, like a ball on a chain [27, 30]. A sequence of three amino acids, the IFM particle on the linker between domain D III and D IV facing the cytoplasmic side of the channel (Fig. 1b), has been identified as the molecular mechanism for "fast" inactivation [31]. Thus "fast" inactivation may function like a "lid" plugging the pore by binding to sites situated on or near the inner vestibule. The role of the IFM particle in binding of local anaesthetics is not fully understood but it seems that this "lid" retains open-channel blockers inside the pore during use-dependent blockade [13, 32]. This situation would mean that the Na⁺ ions could no longer pass through the pore even though the pore is open at the outer mouth. The fast inactivated state seems to play an important role in high affinity local anaesthetic binding (as discussed below); furthermore, the movement of the S4 segment in gating (important for activation and closing as described above) directly influences fast inactivation and *vice versa* [33].

Mechanisms of block

QX 314 is a quaternary derivative of lidocaine with a permanent positive charge. It is not in clinical use but has shown interesting features which have helped to reveal the mechanism of Na^+ channel block. This drug blocks only when applied to the internal side of the neurone or nerve [34]. Nearly all amine local anaesthetic compounds are charged at a pH below 6 except benzocaine. In the uncharged form, the compounds are lipid-soluble [30]. Biophysical calculations of the electric field between the membrane from a blocking model revealed the binding site to be 0.6 parts from the external membrane [35]; this is very close to what is now well known based on the molecular structure of the Na^+ channel (see section below). Biophysical experiments provided the first ideas about blocking a pore at a receptor site with the charged form acting on the "receptor" and required drugs to pass through the lipid membrane to act. Early biophysical experiments revealed that while the first depolarising impulse produced a nearly full-sized Na^+ current in the presence of the local anaesthetics, subsequent impulses elicited smaller and smaller currents [36]. This finding was referred to as "use-dependent block" or "phasic block" [37, 38]. This term describes the phenomenon of progressively developing inhibition during repetitive activation. However, the term is often misunderstood implying a mechanism depending on use of the channel. The underlying mechanism or mechanisms does not necessarily imply the involvement of a binding site within the pore that is only accessible during opening of the channel. This was the basis of the "guarded receptor hypothesis" [39] which proposed that the receptor site for local anaesthetics is protected within the pore and requires it to be open to produce an effect. But the later "modulated receptor model" suggested that a drug, namely local anaesthetics, modify the behaviour of the channel due to preferential affinity towards a conformational state [40]. The preferred state underlying phasic blockade is the inactivated state (mostly the fast) which explain the measured modifications of the rate constants of gating [38, 41, 42].

It followed from use-dependent block of channels that at higher firing frequencies of nerve fibres, that lower concentrations of local anaesthetics are needed to reduce the evoked firing frequencies but higher concentrations are required to produce complete blockade (Fig. 3) [10]. This phenomenon might explain the clinical observed partial blockade of sensory qualities at the begin or end of a local anaesthesia. The overwhelming majority of studies have been performed on tetrodotoxin-sensitive Na^+ channels, which form the main depolarising conductance on fast conducting fibres [10, 38, 43]. However, the tetrodotoxin-insensitive Na^+ channels, mainly found in the peripheral nervous system in smaller neurons and slow conducting fibres, reveal a similar behaviour in terms of phasic blockade except that higher concentrations of local anaesthetics are required [41, 44]. This effect on tetrodotoxin-insensitive Na$^+$ channels it explains the impact of local anaesthetics on action potentials insensitive to tetrodotoxin in small sensory neurons, and the partial reduction of C-fibre compound action potentials (Figs. 2 and 3) [45, 46].

Molecular determinants of local anaesthetic action on Na$^+$ channels

In the last decade it was possible to define the "binding receptor" for local anaesthetics with the knowledge of the primary structure of pore-forming α-subunit of the Na$^+$ channels. Experiments on the rat brain Na$^+$ channel IIa (Na$_V$1.2) revealed that the channel could be made virtually insensitive to use-dependent block by exchanging the amino acid phenylalanine (F) at position 1764 with alanine (A; referred to as F1764A in D IV). Thus while the halfmaximal inhibiting concentration (IC$_{50}$) of anaesthetics at the wild type Na$^+$ channel inactivated with depolarising prepulses are 300-fold lower than that required for tonic blockade, for F1764A mutant channels the IC$_{50}$ values for tonic inhibition are three-fold greater than for wild type Na$^+$ channels, and the IC$_{50}$ values in inactivated Na$^+$ channels are only six times smaller than for tonic blockade. The mutant Y1771A exhibited less use-dependent block and reduced drug binding at depolarised potentials, but the effect was smaller than for F1764A [33]. These results led to a model for the "receptor" binding site of local anaesthetics in the pore of the Na$^+$ channel. The residues F1764A and Y1771A described above are hydrophobic aromatic residues separated by two turns on the same face of the α- helix of the pore-forming S6 segment (Fig. 1b). These amino acids are about 11 Å apart which corresponds well with the observation that a local anaesthetic molecule between 10–15 Å in length is required to induce channel blockade. Local anaesthetics have positively charged moieties at either end which could interact through hydrophobic or π-electrons with these hydrophobic amino acid residues [47, 48]. Substitution of these residues with alanine changed the size and the chemical properties but with little effect on its secondary structure [49]. These alanine substitutions revealed a 10–100-fold reduction in the affinity of local anaesthetics for the open and inactivated states of the channels suggesting that these residues modulate the extracellular access of local anaesthetics to their binding site. This suggestion is supported by experiments with QX314, the permanently charged local anaesthetic, which showed a use-dependent block of more than 50% in the wild-type Na$^+$ channel but produced nearly no use-dependent block in the F1764A mutant. In these types of experiments the only possible access route to this site is through the open channel; recovery times comparable to those in the wild-type indicate that the escape pathway of the local anaesthetic is not altered.

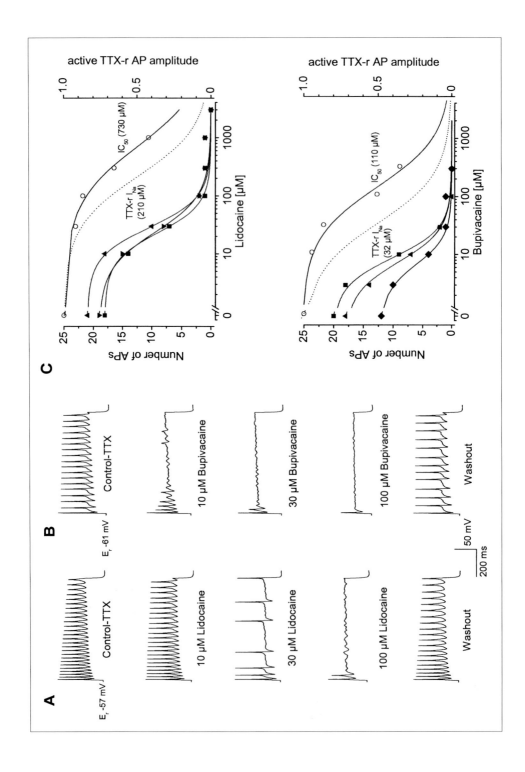

Another mutation oriented towards the pore, I1760A (Fig. 1b), did not alter the local anaesthetic affinity of either the open or the inactivated state of the Na⁺ channel [33]. But it is worth noting that the rate of drug dissociation from the mutant channel was found to be eight times faster than for wild-type channels. Because the I1760A mutation is close to the extracellular side of the S6 segment at the channel mouth, and isoleucine is a bulky residue, it seems likely that the drugs could escape more easily from the mutated channel. Indeed QX314, usually ineffective when applied externally, blocked rapidly this mutant suggesting that the mutation created a pathway for the extracellular drug to enter. Such a structural modification could provide mechanisms for faster recovery from the drug-bound state but still have nearly unchanged use-dependent block properties [33].

The emerging picture provides for two binding sites in the pore of the brain Na⁺ channel (position 1764 and 1771) whose hydrophobic portions interact with the local anaesthetic molecule. The residue oriented more towards the mouth of the pore (1760) guards the fast escape of the drug molecule to the extracellular side and protects the channel from extracellular drugs.

It should be noted that the following mutations in the S6 segment, I1761A, V1767A and N1769A, cause more pronounced blockade at holding potentials of −90 mV. The inactivation curve for these mutants was found to be shifted by

Figure 3
Reduction of firing frequency of TTX-resistant action potentials by application of lidocaine and bupivacaine in a thin slice preparation of rat dorsal root ganglion. A: trains of TTX-resistant action potentials blocked by increasing concentrations of lidocaine. Oscillations can be observed after the action potential at 100 μM lidocaine. Note that the remaining single action potential at the beginning of the current injection is only slightly diminished. B: showing reduction of firing frequency and number of TTX-resistant action potentials by bupivacaine, in another neuron. Oscillations after the AP are mainly at 10 and 30 μM bupivacaine, 750 ms current stimuli of 400 pA. C: concentration-effect-curve for the number of TTX-resistant action potentials and blockade of active amplitudes of TTX-resistant action potentials. A Michaelis-Menten equation was fitted to the data revealing IC$_{50}$ values of 24, 27 and 23 μM for 300 (squares), 400 (triangles) and 500 pA (inverted triangles) current stimuli, respectively, from a single neuron. Blockade of active amplitudes of TTX-resistant APs from the same neuron (open circles, right ordinate) revealed a half-maximal concentration (IC$_{50}$) of 730 μM lidocaine. For comparison the reduction of TTX-resistant Na⁺ current (dotted line; from [41]) is given with an IC$_{50}$ of 210 μM lidocaine. Reduction of numbers of APs revealed IC$_{50}$ values of 7, 9 and 7 μM bupivacaine for 200 (diamonds), 300 (squares) and 400 pA (triangles) stimuli, respectively. IC$_{50}$ of reduction of active TTX-resistant AP amplitudes (open circles, same neuron) was 110 μM bupivacaine. IC$_{50}$ of reduction of TTX-resistant Na⁺ current (dotted line) was 32 μM bupivacaine (adapted from [45] with permission from Pain).

7–13 mV to more negative potentials; it is proposed that the increased sensitivity to local anaesthetics might be caused by an increased proportion of inactivated Na^+ channels. These residues (1761, 1766, and 1769) are oriented away from the inner pore towards the lipid bilayer of the membrane. One possibility may be that mutations to these residues increase sensitivity of the inner pore of the local anaesthetic receptor by allosteric mechanisms. These findings can also explain the higher affinity of lipophilic local anaesthetics under resting conditions. Furthermore these mutations augment the inactivation of the Na^+ channels, perhaps by inducing the binding site in the pore in functionally inactivated binding conformation [33].

Corresponding results have been found at structurally similar amino acid positions for heart and muscle Na^+ channels. This might explain why no substantial difference in affinity to local anaesthetics is found between the different Na^+ channel types and thus no selectivity could be detected between the Na^+ channel subtypes [32]. Only the position numbers of the amino acids are different, e.g., 1579 and 1586 in rat skeletal muscle Na^+ channel ($Na_V1.4$) correspond with position 1764 and 1771 in rat brain Na^+ channels ($Na_V1.2$).

D III appears also important in binding of local anaesthetics, with similar observations to those described above for D IV [21, 50]. Point mutations in the S6 segment in D III reduces affinity of local anaesthetics for neuronal and muscle Na^+ channels. This indicates that at least some drug molecules bound primarily at the S6 segment in D IV but the molecule also binds to another part of the channel pore at the S6 segment of D III.

The local anaesthetic benzocaine is the only clinically used local anaesthetic compound which remains uncharged at physiologic pH due to its low pKa. Despite this, benzocaine is thought to share the same binding site within the inner pore as do other local anaesthetics [51]. Yet benzocaine's reported low efficacy ($IC_{50} \sim 800\ \mu M$ for Na^+ currents) and lack of use-dependent block [51, 52] may suggest otherwise.

Only small differences in the efficacy of enantiomers of the order of around 1.5 ($Na_V1.5$) [53] are conferred by mutation of residues of 1760 and 1765 of Na^+ channels from skeletal muscle and human heart cells ($Na_V1.5$, the corresponding position in $Na_V1.2$ is 1764 and 1771). This might explain the low stereo selectivity of local anaesthetics in contrast to other biological active ligands.

General anaesthetics and ion channels

Clearly, clinical effects of general and local anaesthetics are different. However, both classes of anaesthetics have an analgesic effect in common. It was already of interest for earlier research whether typical general anaesthetics – mainly volatile ones – influenced the ion conductances of the peripheral nerve [54–56]. Indeed, a partial blockade of mainly Na^+ currents was found which was not sufficient to suggest peripheral conduction blockade as a general mechanism. General anaesthetics are less effective on voltage-gated ion channels than on ligand-gated ion channels in the central nervous system. Remarkably, subsequent research found that voltage-gated currents, namely Na^+ and K^+ channels, could be influenced effectively at clinically relevant concentrations in certain species [57–59]. For Na^+ channels, it was reported that the half-maximal values of inactivation were shifted to more hyperpolarised values, resulting in a lower availability of Na^+ channels.

Besides voltage gated K^+ channels, a class of leak or background K^+ selective channels is known. These channels are formed by dimers of subunits each containing four transmembrane segments and two conserved P (pore) domains, they are therefore named 2P domain K^+ channels [60]. In this class of 2P domain K^+ channels, a number were reported to be activated by volatile anaesthetics such as TREK-1, TREK-2, TASK-1 and TASK-2. TASK-1 is in contrast to the other types as it is also blocked by the local anaesthetic bupivacaine [61]. The 2P domain K^+ channels are not uniformly affected by volatile anaesthetics; in contrast to TASK-1, TASK-2 is also stimulated by chloroform, while TASK-1 is partially inhibited by diethyl ether [62]. Even though the stimulation of TREK-1, TREK-2, TASK-1 and TASK-2 activity by halothane is specific, other 2P domain K^+ channels – TWIK-2, THIK-1, TALK-1 and TALK-2 – are inhibited [60]. For channels sensitive to anaesthetics, it was demonstrated that the carboxy terminus was crucial and not the amino terminus [62].

Even though the majority of effects by volatile anaesthetics were reported on 2P domain K^+ channels in the central nervous system there might be a contribution in the peripheral nervous system. An earlier report found a voltage-independent background K^+ channel which was highly sensitive to local anaesthetics and which was involved in setting the resting potential in axons [63].

Toxicity

Evidence from both clinical studies and animal models suggests that lidocaine is neurotoxic possibly via a direct action on sensory neurons [64, 65]. It was observed that lidocaine concentrations already at 30 mM induced DRG neuronal death after a 4 min exposure. At these concentrations lidocaine depolarises DRG neurons with an EC_{50} of 14 mM. Remarkably the depolarisation involved voltage-gated Na^+ currents and is associated with increase in the concentration of intracellular Ca^{2+} ions (EC_{50} of 21 mM) *via* Ca^{2+} influx through the plasma membrane as well as release of Ca^{2+} from intracellular stores. The pivotal role played by intracellular calcium is reflected by the finding that lidocaine-induced neurotoxicity is attenuated in Ca^{2+}-free bath solution by preloading neurons with BAPTA, a strong and fast calcium chelator [66]. It can be assumed that other local anaesthetics can act similarly, although the concentrations at which direct neurotoxicity occurs has yet to be established [67].

Perspectives for local anaesthetics

Early hypotheses based on non-specific interactions of lipid-soluble anaesthetics with membrane bilayers have largely given way to the current idea that membrane-associated proteins, particularly ion channels, are specifically modulated by local anaesthetics. Indeed, Na^+ channels have been identified as a major target with two different blocking mechanisms, tonic and phasic. The use-dependent (phasic) block by local anaesthetics seems to be the mechanism that underlies the very high sensitivity of Na^+ channels which is based on the binding of local anaesthetic molecules in the channel pore to few specific amino acids. Frequency-dependent inhibition seems to be a common characteristic of sodium channel blockers: local anaesthetics, anticonvulsants and class I antiarrhythmics. Interestingly the mechanism of pain inhibition in the peripheral nervous system by anticonvulsants and tricyclic antidepressants is suggested to be the frequency-dependent blockade of voltage gated Na^+ channels [68–71]. Therefore it might be one road to success to search for substances with high affinity to the intrinsic receptor at higher frequencies (as in chronic pain) compared to low affinity at resting conditions which might exhibit less side effects.

A recent investigation focused on the goal of testing the benzomorphan derivative crobenetine (BIII 890 CL), which produces very pronounced use-dependence with the phasic blockade being about 2,000 times more potent than its tonic blockade [20]. Even though it was designed to protect the brain after permanent focal cerebral ischemia, its highly use-dependent Na^+ channel block makes it a possible candidate as a local anaesthetic and for treatment of neuropathic pain [72, 73]. Another specific suppression of pain might be expected from a drug that targets TTX-resistant Na^+ channels (e.g., $Na_V1.8$ and $Na_V1.9$), whose expression is confined to Aδ- and C-fibres mediating pain. However, the development of drugs which exhibit selective blockade of neuronal TTX-resistant Na^+ channels while leaving TTX-sensitive channels unblocked has yet to be accomplished.

References

1   Noda M, Ikeda T, Suzuki H, Takeshima H, Takahashi T, Kuno M, Numa S (1986) Expression of functional sodium channels from cloned cDNA. *Nature* 322: 826–828
2   Freud S (1884) Über Coca. *Zentralbl Ges Ther* 2: 289–314
3   Corning JL (1885) Spinal anaesthesia and local medication of the cord. *NY Med J* 42: 183–185
4   Quincke H (1891) Die Lumbalpunktion des Hydrozephalus. *Berl Klin Wochenschr* 28: 929–933
5   Bier A (1899) Versuche über Cocainisirung des Rückenmarkes. *Dtsch Zeitschr f Chir* 51: 361–369
6   Einhorn A (1899) On the chemistry of local anesthetics. *Munch Med Wochenschr* 46: 1218–1220

7   Gasser HS, Erlanger J (1929) Role of fibre size in the establishment of nerve block by pressure or cocaine. *Am J Physiol* 88: 581–591

8   Gissen AJ, Covino BG, Gregus J (1980) Differential sensitivities of mammalian nerve fibers to local anesthetic agents. *Anesthesiology* 53: 467–474

9   Wildsmith JA, Gissen AJ, Takman B, Covino BG (1987) Differential nerve blockade: esters v. amides and the influence of pKa. *Br J Anaesth* 59: 379–384

10 Raymond SA, Gissen AJ (1987) Mechanisms of differential nerve block. In: GR Strichartz (ed): *Local Anesthetics*, 1st ed. Springer-VerlagBerlin, Heidelberg, 95–165

11 Agnew WS, Levinson SR, Brabson JS, Raftery MA (1978) Purification of the tetrodotoxin-in-binding component associated with the voltage-sensitive sodium channel from Electrophorus electricus electroplax membranes. *Proc Natl Acad Sci USA* 75: 2606–2610

12 Noda M (1993) Structure and function of sodium channels. In: Molecular basis of ion channels and receptors involved in nerve excitation, synaptic transmission and muscle contraction. *Ann NY Acad Sci* 707: 20–37

13 Catterall WA (2000) From ionic currents to molecular mechanisms: the structure and function of voltage-gated sodium channels. *Neuron* 26: 13–25

14 Goldin AL, Barchi RL, Caldwell JH, Hofmann F, Howe JR, Hunter JC, Kallen RG, Mandel G, Meisler MH, Netter YB et al (2000) Nomenclature of voltage-gated sodium channels. *Neuron* 28: 365–368

15 Vassilev PM, Scheuer T, Catterall WA (1988) Identification of an intracellular peptide segment involved in sodium channel inactivation. *Science* 241: 1658–1661

16 Terlau H, Heinemann SH, Stuhmer W, Pusch M, Conti F, Imoto K, Numa S (1991) Mapping the site of block by tetrodotoxin and saxitoxin of sodium channel II. *FEBS Lett* 293: 93–96

17 Shon KJ, Olivera BM, Watkins M, Jacobsen RB, Gray WR, Floresca CZ, Cruz LJ, Hillyard DR, Brink A, Terlau H et al (1998) mu-Conotoxin PIIIA, a new peptide for discriminating among tetrodotoxin-sensitive Na channel subtypes. *J Neurosci* 18: 4473–4481

18 Wang SY, Wang GK (1999) Batrachotoxin-resistant Na⁺ channels derived from point mutations in transmembrane segment D4-S6. *Biophys J* 76: 3141–3149

19 Wang SY, Barile M, Wang GK (2001) Disparate role of Na⁺ channel D2-S6 residues in batrachotoxin and local anesthetic action. *Mol Pharmacol* 59: 1100–1107

20 Carter AJ, Grauert M, Pschorn U, Bechtel WD, Bartmann-Lindholm C, Qu YS, Scheuer T, Catterall WA, Weiser T (2000) Potent blockade of sodium channels and protection of brain tissue from ischemia by BIII890CL. *Proc Natl Acad Sci USA* 97: 4944–4949

21 Wang SY, Nau C, Wang GK (2000) Residues in Na⁺ channel D3-S6 segment modulate both batrachotoxin and local anesthetic affinities. *Biophys J* 79: 1379–1387

22 Isom LL, De Jongh KS, Patton DE, Reber BF, Offord J, Charbonneau H, Walsh K, Goldin AL, Catterall WA (1992) Primary structure and functional expression of the β1 subunit of the rat brain sodium channel. *Science* 256: 839–842

23 Baker MD, Wood JN (2001) Involvement of Na⁺ channels in pain pathways. *Trends Pharmacol Sci* 22: 27–31

24 Malhotra JD, Kazen-Gillespie K, Hortsch M, Isom LL (2000) Sodium channel β sub-

units mediate homophilic cell adhesion and recruit ankyrin to points of cell-cell contact. *J Biol Chem* 275: 11383–11388

25 Benitah JP, Chen Z, Balser JR, Tomaselli GF, Marban E (1999) Molecular dynamics of the sodium channel pore vary with gating: interactions between P-segment motions and inactivation. *J Neurosci* 19: 1577–1585

26 Hodgkin AL, Huxley AF (1952) A quantitative description of membrane current and its application to conduction and excitation in nerve. *J Physiol* 117: 400–544

27 Bezanilla F, Armstrong CM (1977) Inactivation of the sodium channel. I. Sodium current experiments. *J Gen Physiol* 70: 549–566

28 Loots E, Isacoff EY (2000) Molecular coupling of S4 to a K⁺ channel's slow inactivation gate. *J Gen Physiol* 116: 623–636

29 Perozo E, Cortes DM, Cuello LG (1999) Structural rearrangements underlying K⁺-channel activation gating. *Science* 285: 73–78

30 Hille B (ed) (2001) *Ion channels of excitable membranes*. 4th edition. Sinauer Associates, Inc. Sunderland, Massachusetts

31 Eaholtz G, Scheuer T, Catterall WA (1994) Restoration of inactivation and block of open sodium channels by an inactivation gate peptide. *Neuron* 12: 1041–1048

32 Balser JR (2001) The cardiac sodium channel: gating function and molecular pharmacology. *J Mol Cell Cardiol* 33: 599–613

33 Ragsdale DS, McPhee JC, Scheuer T, Catterall WA (1994) Molecular determinants of state-dependent block of Na⁺ channels by local anesthetics. *Science* 265: 1724–1728

34 Frazier DT, Narahashi T, Yamada M (1970) The site of action and active form of local anesthetics. II. Experiments with quaternary compounds. *J Pharmacol Exp Ther* 171: 45–51

35 Woodhull AM (1973) Ionic blockage of sodium channels in nerve. *J Gen Physiol* 61: 687–708

36 Strichartz GR (1973) The inhibition of sodium currents in myelinated nerve by quaternary derivatives of lidocaine. *J Gen Physiol* 62: 37–57

37 Courtney KR (1975) Mechanism of frequency-dependent inhibition of sodium currents in frog myelinated nerve by the lidocaine derivative GEA. *J Pharmacol Exp Ther* 195: 225–236

38 Ulbricht W (1981) Kinetics of drug action and equilibrium results at the node of Ranvier. *Physiol Rev* 61: 785–828

39 Starmer CF, Grant AO, Strauss HC (1984) Mechanisms of use-dependent block of sodium channels in excitable membranes by local anesthetics. *Biophys J* 46: 15–27

40 Hille B (1977) Local anesthetics: hydrophilic and hydrophobic pathways for the drug-receptor reaction. *J Gen Physiol* 69: 497–515

41 Scholz A, Kuboyama N, Hempelmann G and Vogel W (1998) Complex blockade of TTX-resistant Na⁺ currents by lidocaine and bupivacaine reduce firing frequency in DRG neurons. *J Neurophysiol* 79: 1746–1754

42 Kendig JJ, Courtney KR, Cohen EN (1979) Anesthetics: molecular correlates of voltage-

and frequency-dependent sodium channel block in nerve. *J Pharmacol Exp Ther* 210: 446–452

43　Scholz A (2002) Mechanisms of (local) anaesthetics on voltage-gated sodium and other ion channels. *Br J Anaesth* 89: 52–61

44　Roy ML, Narahashi T (1992) Differential properties of tetrodotoxin-sensitive and tetrodotoxin-resistant sodium channels in rat dorsal root ganglion neurons. *J Neurosci* 12: 2104–2111

45　Scholz A, Vogel W (2000) Tetrodotoxin-resistant action potentials in dorsal root ganglion neurons are blocked by local anesthetics. *Pain* 89: 47–52

46　Strassman AM, Raymond SA (1999) Electrophysiological evidence for tetrodotoxin-resistant sodium channels in slowly conducting dural sensory fibers. *J Neurophysiol* 81: 413–424

47　Bokesch PM, Post C, Strichartz G (1986) Structure-activity relationship of lidocaine homologs producing tonic and frequency-dependent impulse blockade in nerve. *J Pharmacol Exp Ther* 237: 773–781

48　Heginbotham L, Abramson T, MacKinnon R (1992) A functional connection between the pores of distantly related ion channels as revealed by mutant K$^+$ channels. *Science* 258: 1152–1155

49　Richardson JS (1981) The anatomy and taxonomy of protein structure. *Adv Protein Chem* 34: 167–339

50　Yarov-Yarovoy V, Brown J, Sharp EM, Clare JJ, Scheuer T and Catterall WA (2001) Molecular determinants of voltage-dependent gating and binding of pore-blocking drugs in transmembrane segment IIIS6 of the Na$^+$ channel α subunit. *J Biol Chem* 276: 20–27

51　Schmidtmayer J, Ulbricht W (1980) Interaction of lidocaine and benzocaine in blocking sodium channels. *Pflügers Arch* 387: 47–54

52　Wang GK, Quan C, Wang S (1998) A common local anesthetic receptor for benzocaine and etidocaine in voltage-gated mu1 Na$^+$ channels. *Pflügers Arch* 435: 293–302

53　Nau C, Wang SY, Strichartz GR, Wang GK (2000) Block of human heart hH1 sodium channels by the enantiomers of bupivacaine. *Anesthesiology* 93: 1022–1033

54　Haydon DA, Urban BW (1983) The effects of some inhalation anaesthetics on the sodium current of the squid giant axon. *J Physiol* 341: 429–439

55　Butterworth JF, Raymond SA, Roscoe RF (1989) Effects of halothane and enflurane on firing threshold of frog myelinated axons. *J Physiol* 411: 493–516

56　Bean BP, Shrager P, Goldstein DA (1981) Modification of sodium and potassium channel gating kinetics by ether and halothane. *J Gen Physiol* 77: 233–253

57　Friederich P, Benzenberg D, Trellakis S, Urban BW (2001) Interaction of volatile anesthetics with human Kv channels in relation to clinical concentrations. *Anesthesiology* 95: 954–958

58　Frenkel C, Wartenberg HC, Rehberg B, Urban BW (1997) Interactions of ethanol with single human brain sodium channels. *Neurosci Res Commun* 20: 113–120

59　Scholz A, Appel N, Vogel W (1998) Two types of TTX-resistant and one TTX-sensitive

Na⁺ channel in rat dorsal root ganglion neurons and their blockade by halothane. Eur J Neurosci 10: 2547–2556

60   Patel AJ, Honore E (2001) Properties and modulation of mammalian 2P domain K⁺ channels. *TINS* 24: 339–346

61   Buckler KJ, Williams BA, Honore E (2000) An oxygen-, acid- and anaesthetic-sensitive TASK-like background potassium channel in rat arterial chemoreceptor cells. *J Physiol* 525 Pt 1: 135–142

62   Patel AJ, Honore E, Lesage F, Fink M, Romey G, Lazdunski M (1999) Inhalational anesthetics activate two-pore-domain background K⁺ channels. *Nat Neurosci* 2: 422–426

63   Brau ME, Nau C, Hempelmann G, Vogel W (1995) Local anesthetics potently block a potential insensitive potassium channel in myelinated nerve. *J Gen Physiol* 105: 485–505

64   Gold MS, Reichling DB, Hampl KF, Drasner K, Levine JD (1998) Lidocaine toxicity in primary afferent neurons from the rat. *J Pharmacol Exp Ther* 285: 413–421

65   Johnson ME (2000) Potential neurotoxicity of spinal anesthesia with lidocaine. *Mayo Clin Proc* 75: 921–932

66   Johnson ME, Saenz JA, DaSilva AD, Uhl CB, Gores GJ (2002) Effect of local anaesthetic on neuronal cytoplasmic calcium and plasma membrane lysis (necrosis) in a cell culture model. *Anesthesiology* 97: 1466–1476

67   Kuboyama N, Nakoa S, Moriya Y, Scholz A, Vogel W (1997) Bupivacaine-included Ca²⁺ release on intracellular Ca²⁺ stores in rat DRG neurones. *Jpn J Pharmacol* 73: O103–104

68   Mike A, Karoly R, Vizi ES, Kiss JP (2004) A novel modulatory mechanism of sodium currents: frequency-dependence without state-dependent binding. *Neuroscience* 125: 1019–1028

69   Nau C, Seaver M, Wang SY, Wang GK (2000) Block of human heart hH1 sodium channels by amitriptyline. *J Pharmacol Exp Ther* 292: 1015–1023

70   Wang GK, Russell C, Wang SY (2004) State-dependent block of voltage-gated Na(⁺) channels by amitriptyline via the local anesthetic receptor and its implication for neuropathic pain. *Pain* 110: 166–174

71   Barnet CS, Tse JY, Kohane DS (2004) Site 1 sodium channel blockers prolong the duration of sciatic nerve blockade from tricyclic antidepressants. *Pain* 110: 432–438

72   Krause U, Weiser T, Carter AJ, Grauert M, Vogel W, Scholz A (2000) Potent, use-dependent blockade of TTX-resistant Na⁺-channels in dorsal root ganglion neurones by BIII 890 CL. *Eur J Neurosci* 12: S387

73   Laird JMA, Carter AJ, Grauert M, Cervero F (2002) Analgesic activity of a novel use-dependent sodium channel blocker, crobenetine, in mono-arthritic rats. *Br J Pharmacol* 134: 1742–1748

74   Sato C, Ueno Y, Asai K, Takahashi K, Sato M, Engel A, Fujiyoshi Y (2001) The voltage-sensitive sodium channel is a bell-shaped molecule with several cavities. *Nature* 409: 1047–1051

75   Catterall WA (2001) Structural biology – A 3D view of sodium channels. *Nature* 409: 988–991

Future directions in sodium channel research

John N. Wood

Molecular Nociception Group, Biology Department, University College London (UCL), Gower Street, London WC1E 6BT, UK

Introduction

Although the human genome comprises approximately 25,000 genes, the number of functional proteins expressed may be massively amplified by the production of splice variants. For example, the single *Drosophila* gene Dscam has 38,016 possible alternatively spliced forms [1]. Most estimates suggest that more than 40% of mammalian genes are alternatively spliced, and voltage-gated sodium channels are members of this set. From the earliest molecular characterisation of sodium channels, it has been clear that different isoforms may be expressed in development.

Studies in insects have given insights into the regulation of expression of sodium channel genes and provided clear evidence that alternatively spliced isoforms may have distinct functional properties. Insect sodium channels are instructive because they have been shown to undergo developmental regulation of splice variants, RNA editing and to display functional diversity of splice variants. Such events may also occur in mammals. Specific pharmacological manipulation of sodium channel isoforms is clearly a challenging route to understanding function. This chapter focuses on the genetics of sodium channel expression, and how manipulating gene expression in transgenic mice will continue to provide useful insights into the specialised roles of sodium channel subtypes.

Transcriptional regulation of sodium channel expression

The cell-specific pattern of expression of different sodium channels shown in Figure 1 demonstrates the exquisite subtlety of transcriptional regulatory mechanism that control channel expression. Although we do not have a comprehensive knowledge of any sodium channel promoter's structure and association with a particular transcription factors, we do have a number of insights into some aspects of sodium channel regulatory motifs. A short sequence found upstream of neuronal sodium channel genes (as well as other neuronal genes) was identified and named NRSE

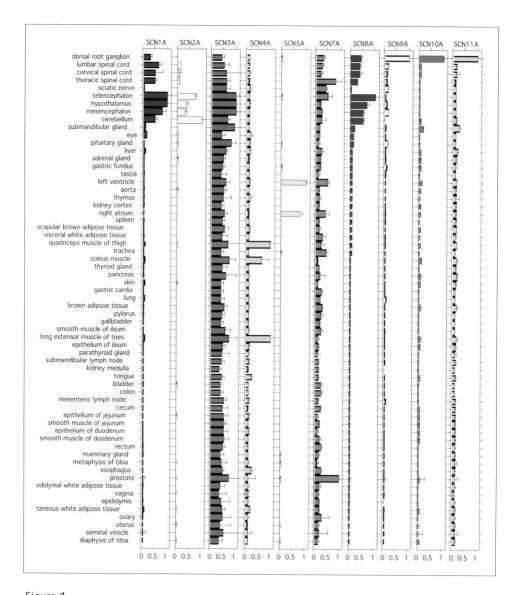

Figure 1
Sodium channel diversity in different tissues
Taqman analysis of the relative levels of expression of sodium channel α-subunits in cyno-
molgus monkeys – reproduced with permission from [10].

(Neuron restricted silencing element) or RE-1 (repressor element 1) [2, 3]. Transcription factors that bound to the motif were found to act as inhibitors of gene
expression in non-neuronal cells. These proteins were named REST (RE-1 silencing

transcription factor), or NRSF (Neuron-restrictive silencer factor) and mutational analysis identified a single zinc finger motif in the carboxyl-terminal domain of the factor as essential for repressing type II sodium channel-derived reporter genes [4]. Intriguingly, other proteins recruited to the NRSF complex have been shown to exert repressive actions on gene expression over areas of chromatin adjacent to the NRSE sequence [5]. The inhibitory activity of the complex can be further modulated by double stranded RNA molecules that have the same sequence as NRSE/RE-1, and are found in developing neuronal precursors. These regulatory RNA molecules are able to switch the repressor function of the complex to an activator role [6]. In this way, the assumption of a neuronal phenotype seems to depend in part upon regulatory RNAs driving gene expression downstream of NRSE/RE1 motifs. Sodium channels are known to be expressed at the very earliest stages of the appearance of a neuronal phenotype in the mouse. These studies highlight the significance of sodium channel expression in neuronal function throughout development [7].

Regulation of expression of muscle sodium channel forms has also been found to involve both inducing and silencing elements. Muscle-associated sodium channel genes (SKm2 or $Na_V1.4$) have a number of MyoD transcription factor binding E-box motifs upstream of the structural gene, as well as a muscle-restrictive enhancer element (MRSE) at least 2 kb upstream from the core promoter [8].

Apart from the tissue-specific control of sodium channel expression most obviously demonstrated by the presence of neuronal and muscle isoforms, there is evidence that the relative levels of sodium channel transcripts vary in development. Early studies of the developing rat gave us the first indication that the type III sodium channel $Na_V1.3$ is prevalent in rat embryos and expressed at much lower levels in adult neuronal tissues, whilst the Type 1 and 2 channels $Na_V1.1$ and $Na_V1.2$ are expressed in variable patterns in adult tissues [9]. Interestingly, this pattern of expression does not seem to hold true in cynomolgus monkeys where $Na_V1.3$ is broadly expressed, albeit at low levels in both central and peripheral tissues in the adult [10]. Thus the pattern of expression of human sodium channels may vary markedly from that described in detail in rodents. Felts et al. [11] extended the rat development analysis with probes against $Na_V1.1$, 1.2, 1.3, 1.6, and 1.7 and showed a complex pattern of developmentally regulated functional channel expression in both peripheral and central neurons.

As with mammals, developmental regulation of sodium channel genes also occurs in *Drosophila*. Two genes, Para and DSC1, encode functional sodium channel α-subunits that are expressed in distinct patterns during development [12]. Para seems to be the major neuronal form of sodium channels, and its expression is upregulated by an associated trans-membrane protein TipE, that is required to avoid adult paralysis [13]. The Para channel seems to be expressed throughout development, while DSC-1 is expressed later and shows an overlapping pattern of expression with Para in pupal and adult flies.

Accessory subunits that are assumed to be uniquely associated with voltage-gated sodium channels also show developmentally-regulated patterns of expression. Shah et al. [14] showed that in the developing rat, the β3-subunit was prevalent, and this subunit remained expressed in adult hippocampus and striatum. β1- and β2-subunits were expressed after postnatal day 3 in the rat in central and peripheral tissues. The developmental pattern of expression pf the β2-like subunit β4 [15] that is also expressed both centrally and peripherally has yet to be described. The functional significance of β-subunit expression for sodium channel kinetics properties and their tethering to extracellular signalling molecules has been explored extensively (Okuse and Baker, this volume, and [16]).

Splice variants of sodium channels

A further level of complexity in the expression of sodium channels is created by the existence of splice variants of sodium channel α-subunits that may be regulated during development and by regulators of splicing choice in the adult. Because fly genetics is so advanced, more information is available about *Drosophila* splice choice sodium channel variants and their functional roles than the vertebrate equivalents. However, it seems reasonable to suppose that alternative splicing and RNA editing and transport may also have roles in regulating mammalian sodium channel function. In *Drosophila*, a mutation in a double-stranded RNA helicase led to a lowering of expression of Para-encoded sodium channels. Reenan et al. [17] showed that this was due to a failure to edit the Para transcript with an adenosine to inosine substitution, which apparently required the helicase for secondary structure modification of the mRNA transcript. At least three positions in the Para transcript are known to be edited [18] by adenosine deaminase to give A to I substitutions and these events are developmentally regulated. The editing process requires a complementary sequence of intronic RNA to form secondary structure with the edited sequence in a similar manner to that demonstrated for mammalian glutamate receptors [19]. Other editing events in cockroach sodium channels and *Drosophila* Para have been correlated with dramatic functional changes. Liu et al. [20] have shown that a U to C editing event resulting in a phenylalanine to serine modification can produce a sodium channel with persistent tetrodotoxin-sensitive properties rather similar to currents identified in mammalian CNS neurons, raising the possibility that similar events could occur in mammals. Song et al. [21] have catalogued further editing events that have functional consequences in terms of thresholds of activation and are developmentally regulated in the cockroach. Tan et al. [22] have also found that alternatively spliced transcripts can have distinct pharmacological profiles as well as altered gating characteristics. They found alternative exons encoding transmembrane segments in domain 3 of a cockroach sodium channel, which had conserved splice sites across evolution in fish, flies, mice and men. The alternatively spliced forms were found in different tissues. One form with a premature stop codon occurred only in the peripheral nervous system while the two other functional forms differed in their sensitivity to pyrethroid insecticides such as deltamethrin. Remarkably, foetal mouse brain also contains transcripts of the SCN8A gene ($Na_V1.6$) that contains a stop codon at the same site as the fly genes, predicting the production of two domain truncated sodium channel transcript [23]. The role of these transcripts is unknown.

In mammals, mutually exclusive exon usage also occurs. The $Na_V1.3$ channel exists as an embryonic or adult spliced form with different exons that code for the S3 and S4 segments in domain 1 of the rat channel. Despite the fact that the two exons both encode 29 amino acids, only a single amino acid residue is altered in these two alternative forms [24]. Single amino acid changes may also occur through alternative 3' splice site selection. Kerr et al. [25] have found that both $Na_V1.8$ and $Na_V1.5$ – two terodotoxin-resistant (TTX-r) sodium channels found in peripheral neurons and the heart respectively – both exist as alternative forms containing an additional glutamine reside within the cytoplasmic loop linking domains 2 and 3 of these channels. As well as alternative exon usage or amino acid insertions, transcripts encoding exon repeats have been identified in dorsal root ganglion neurons. The presence of a transcript with a 3-exon repeat encoding $Na_V1.8$ is enhanced by treatment with NGF, suggesting that this neurotrophin may regulate trans-splicing events in these cells [26]. Once again, the functional consequences of these conserved changes have yet to be established.

Raymond et al. [10] carried out a comprehensive analysis of the expression of sodium channel isoforms in all tissues of the cynomolgus monkey, and also assessed the expression of splice variants of $Na_V1.6$, 1.7 and 1.9. Some splice variants of $Na_V1.6$ were only expressed in peripheral sensory neurons, and were regulated by damage to these neurons. These primate data are of particular interest, as the expression pattern is likely to be similar to that found in man.

Muscle sodium channels have also been found to exist in alternatively spliced or polymorphic forms in man [27, 28]. The TTX-resistant channel associated with cardiac function $Na_V1.5$ contains an H558R polymorphism in 30% of subjects; while a glutamine deletion at position 1077 (see [25]) occurred in 65% of cases. Subtle functional differences concerning peak currents were described for these variants [28]. The skeletal muscle form $Na_V1.4$ was also found in cardiac tissue [27]. Conversely PCR and *in situ* hybridisation data have now demonstrated that the cardiac channel is found in the central nervous system [29, 30]. Interestingly, mutations in the cardiac channel that causes Brugada sudden death syndrome may also cause epileptic episodes consistent with this distribution of expression [31].

The regulation of splice choice in response to external signals is still little understood. Buchner et al. [32], studying a modifier locus in different mouse lines that determines the lethality of a $Na_V1.6$ splice site mutation, discovered that the efficiency of action of a splice factor determined the amount of function channel produced and hence the lethality of the original mutation. Thus on a C57Bl6 background little correctly spliced mRNA (5%) was produced causing a lethal phenotype, whilst on a wild-type background, 10% of the transcripts were correctly spliced leading to a viable, if dystonic, phenotype.

Novel roles for sodium channels

Recent papers suggest roles for sodium channels in regulating synaptic efficacy, as well as functions in immune system cells. However, intriguing claims that BDNF can depolarise hippocampal neurons as effectively as glutamate have as yet not been replicated by other groups. Similarly, claims that BDNF can activate $Na_V1.9$ over a millisecond timescale have not been confirmed [33, 34].

Attention has focused on the role of sodium channels in nerve and muscle, although it has long been know that sodium channels are also present in non-excitable cells such as Schwann cells and glia, including microglia, as well as lymphocytes [35]. Recent work has demonstrated a potentially important role in immune system function. Macrophages and microglia express $Na_V1.6$, a channel that is broadly expressed in the nervous system and which is functionally compromised in the naturally occurring *med* mutant that leads to dystonia. Interestingly, macrophage function is also inhibited in these animals. When microglia or macrophages are activated $Na_V1.6$ expression is upregulated, and this event seems to be important in terms of phagocytic activity, as the uptake of latex beads is partially blocked in *med* macrophages, or in normal macrophages treated with TTX. This suggests that voltage-gated sodium channels play an important hitherto overlooked functional role in immune system cell function [35].

Sodium channel diversity

Given the similar properties of voltage-gated sodium channels, why are there so many different α-subunits, yet alone multiple splice variants? Firstly, the trafficking of different sodium channels to distinct cellular locations (nerve terminals, nodes of Ranvier, etc.) and the regulation of this process may provide a number of options to control neuronal excitability in different physiological contexts. Thus trafficking of $Na_V1.8$ into the cell membrane through its interaction with p11 which is massively upregulated by NGF could play an important role in inflammatory pain [36]. Secondly, different structural features of the channels mean that their response to post-translational modification by enzymes such as protein kinases is quite distinct. Phosphorylation on intracellular serines in $Na_V1.8$ increase peak current density, whilst in other channels present in DRG neurons, for example $Na_V1.7$, peak current diminishes [37, 38]. Primary sequence also determines the repriming characteristics of the different channels as well as their threshold of activation which are crucial determinants of neuronal excitability [39].

Finally, the topological relationship between different sodium channels and ligand-gated channels and receptors involved in intercellular signalling may play a role in altering the electrical responses of neurons to incoming signals. There is evidence that disruption of the cytoskeleton has anti-hyperalgesic effects [40] although molecules other than sodium channels must also be involved in this type of regulation.

Selective blockers of sodium channel subtypes

Natural products first demonstrated the diversity of sodium channel subtypes. Tetrodotoxin and saxitoxin distinguish two subsets of sodium channels in terms of amino acids present in the channel atrium. Other natural products also show a selectivity of action on sodium channel subtypes. The venom from marine predatory cone shell snails (*Conus*) contains a complex cocktail of neurotoxic disulphide-rich peptides, termed conotoxins. Conotoxins are highly selective for different ion channel isoforms, and toxins selective for sodium channels have been identified. The two main classes of conotoxin that affect voltage-gated sodium channels are the μ and δ conotoxins (μ-CTXs and δ-CTXs). The μ-CTXs are a group of homologous peptides belonging to the M superfamily of conotoxins, containing a 6-cysteine residue backbone arranged in a double–single–single–double cysteine order (CC-C-C-CC) forming 3-disulphide loops. μ-conotoxin pIIIA blocks $Na_V1.2$ sodium channels but has little effect on the $Na_V1.7$ channel [41]. Recordings from variant lines of PC12 cells, which selectively express either $Na_V1.2$ or $Na_V1.7$ channel subtypes, verified that the differential block by PIIIA also applied to native sodium current. SmIIIA is the first μ-conotoxin to be identified to block a TTX-r current [42, 43]. SmIIIA irreversibly blocks TTX-R sodium currents recorded from cultured frog dorsal root ganglion (DRG) [42, 43]. Further, SmIIIA irreversibly inhibits frog C-fibre compound action potentials in the presence of TTX. The development of new technology for high-throughput screens of blockers of expressed sodium channels should facilitate the identification of non-peptide selective blockers of different sodium channel subtypes.

Genetic approaches to understanding sodium channel function

The complex regulation of expression of the various sodium channels raises difficulties in understanding their specialised functional roles. It has proved difficult to generate drugs that distinguish between different sodium channel sub-types, and splice variants. Targeting gene deletions to a particular tissue and examining the functional consequences both electrophysiologically and in terms of animal behaviour provides our best hope of unravelling the functional significance of sodium channel diversity. The technology to delete particular exons in both a tissue specific and temporally controlled manner has been successfully developed over the past decade.

Sauer and collaborators [44] exploited the recombinase activity of a bacteriophage enzyme Cre, to delete DNA sequences in mammalian cells that are flanked by lox-P sites. Applying this technology to embryonic stem cells, it has proved possible to generate tissue-specific mouse null mutants. An analogous system exploits the Flp recombinase that recognises Frt sites. By deleting genes only in a subset of cells, it is thus possible to examine the specialised role of a broadly expressed gene in a specific physiological system. Problems of developmental lethality may also be avoided using this approach.

Thus far, mice containing floxed (lox-P flanked) sodium channel genes, $Na_V1.3$, 1.6, and 1.7 have been generated. Such animals are generated after homologous recombination of engineered alleles in embryonic stem cells. The lox-P containing and engineered constructs can be distinguished by southern blots. Similarly the floxed and deleted allele can be distinguished by PCR at both the genomic level and the RNA level (Fig. 2).

There is considerable interest in the specialised role of sodium channel isoforms in nociceptors. In order to ablate genes in sensory ganglia, it is necessary to produce mice in which functional Cre-recombinase is driven by sensory neuron-specific promoters. The effectiveness of expressed Cre in excising lox-P-flanked genes can be measured with a reporter mouse using the β-galactosidase-expressing gene with a floxed (loxP flanked) stop signal. Where Cre removes the stop signal, β-galactosidase activity can be analysed histochemically. The $Na_V1.8$ gene is expressed predominantly in nociceptive sensory neurons, and is completely absent in tissue other than sensory neurons [45, 46]. Heterozygous null mutant $Na_V1.8$ mice are completely normal [45], suggesting that "knocking-in" a Cre-recombinase into the $Na_V1.8$ locus is unlikely to have deleterious effects in heterozygous mice that express single alleles of $Na_V1.8$ and Cre. These mice were constructed and analysed, and showed no phenotypic deficits, while expressing Cre-recombinase in a similar pattern to $Na_V1.8$ [47].

It would be even more useful to generate transgenic mice expressing drug-activatable Cre isoforms exclusively in subsets of sensory neurons. Such an approach would remove the problem of developmental compensatory mechanisms that may mask the phenotype caused by deletion of a particular gene. Tamoxifen-activatable forms of Cre-recombinase has been developed. This form of Cre-recombinase comprises a fusion protein between Cre and a human mutated oestrogen receptor. The addition of tamoxifen, but not endogenous steroids, releases the Cre-recombinase from a cytoplasmic association with HSP90 and allows it to enter the nucleus [48]. This allows the excision of genes at defined periods in adulthood.

Such genetic approaches to understanding channel function are likely to be applied increasingly over the next few years, and together with the use of antisense

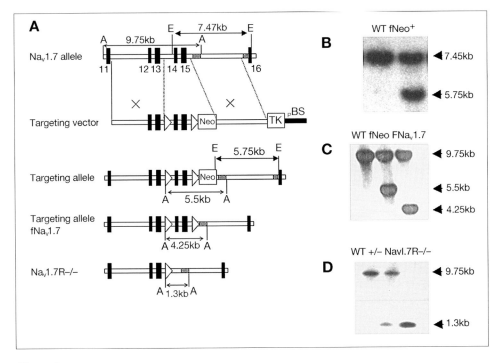

Figure 2
Generation and analysis of lox-P flanked sodium channels for tissue specific deletion of Na$_V$1.7 [47]
Structure of the native Na$_V$1.7 allele, Na$_V$1.7 targeting construct, Floxed fNa$_V$1.7 allele, Floxed fNa$_V$1.7 allele after excision of neo^r and Na$_V$1.7 knockout allele.
b, Southern blot with EcoRI and the external probe confirms correct targeting.
c, Southern blot with ApaI and internal probe confirms the removal of neo^r cassette.
d, Southern blot confirms deletion of exons 14 and 15 in Na$_V$1.7R–/–.
e, PCR was used to detect exon 14–15 deletion in genomic and cDNA from DRG but not spinal cord in Na$_V$1.7R–/–.

RNA and siRNA promise to speed up our understanding of the physiological role of sodium channel splice variants in particular tissues, with useful consequences for the treatment of disease.

Summary

Sodium channels act in concert to play a critical role not only in electrical signalling in the nervous system but also in terms of regulating neuronal excitability in response to external cues. Splice variants, channel editing, post-translational modification and association with accessory proteins all may amplify the repertoire of voltage-gated sodium channels. Using adult inducible knockouts of the various components of sodium channel complexes should provide invaluable information for defining new targets for therapeutic intervention in a much more efficient way than conventional pharmacology allows.

Acknowledgements

We thank the MRC for their long term support.

References

1   Neves G, Zucker J, Daly M, Chess A (2004) Stochastic yet biased expression of multiple Dscam splice variants by individual cells. *Nat Genet* 36: 240–246
2   Schoenherr CJ, Anderson DJ (1995) The neuron-restrictive silencer factor (NRSF): a coordinate repressor of multiple neuron-specific genes. *Science* 267 1360–1365
3   Kraner SD, Chong JA, Tsay HJ, Mandel G (1992) Silencing the type II sodium channel gene: a model for neural-specific gene regulation. *Neuron* 1: 37–44
4   Tapia-Ramirez J, Eggen BJ, Peral-Rubio MJ, Toledo-Aral JJ, Mandel G (1997) A single zinc finger motif in the silencing factor REST represses the neural-specific type II sodium channel promoter. *Proc Natl Acad Sci USA* 94: 1177–1182
5   Lunyak VV, Burgess R, Prefontaine GG, Nelson C, Sze SH, Chenoweth J, Schwartz P, Pevzner PA, Glass C, Mandel G et al (2002) Corepressor-dependent silencing of chromosomal regions encoding neuronal genes. *Science* 298: 1747–1752
6   Kuwabara T, Hsieh J, Nakashima K, Taira K, Gage FH (2004) A small modulatory dsRNA specifies the fate of adult neural stem cells. *Cell* 116(6): 779–793
7   Albrieux M, Platel JC, Dupuis A, Villaz M, Moody WJ (2004) Early expression of sodium channel transcripts and sodium current by cajal-retzius cells in the preplate of the embryonic mouse neocortex. *J Neurosci* 24: 1719–1725
8   Sheng ZH, Zhang H, Barchi RL, Kallen RG (1994) Molecular cloning and functional analysis of the promoter of rat skeletal muscle voltage-sensitive sodium channel subtype 2 (rSkM2): evidence for muscle-specific nuclear protein binding to the core promoter. *DNA Cell Biol* 13: 9–23
9   Beckh S, Noda M, Lubbert H, Numa S (1989) Differential regulation of three sodium channel messenger RNAs in the rat central nervous system during development. *EMBO J* 8: 3611–3616
10   Raymond CK, Castle J, Garrett-Engele P, Armour CD, Kan Z, Tsinoremas N, Johnson JM (2004) Expression of alternatively spliced sodium channel alpha-subunit genes: Unique splicing patterns are observed in dorsal root ganglia. *J Biol Chem* 279: 46234–46241

11   Felts PA, Yokoyama S, Dib-Hajj S, Black JA, Waxman SG (1997) Sodium channel alpha-subunit mRNAs I, II, III, NaG, Na6 and hNE (PN1): different expression patterns in developing rat nervous system. *Brain Res Mol Brain Res* 45: 71–82

12   Hong CS, Ganetzky B (1994) Spatial and temporal expression patterns of two sodium channel genes in Drosophila. *J Neurosci* 14: 5160–5169

13   Feng G, Deak P, Chopra M, Hall LM (1995) Cloning and functional analysis of TipE, a novel membrane protein that enhances Drosophila para sodium channel function. *Cell* 82: 1001–1011

14   Shah BS, Stevens EB, Pinnock RD, Dixon AK, Lee K (2001) Developmental expression of the novel voltage-gated sodium channel auxiliary subunit beta3, in rat CNS. *J Physiol* 534 763–776

15   Yu FH, Westenbroek RE, Silos-Santiago I, McCormick KA, Lawson D, Ge P, Ferriera H, Lilly J, DiStefano PS, Catterall WA et al (2003) Sodium channel beta4, a new disulfide-linked auxiliary subunit with similarity to beta2. *J Neurosci* 23: 7577–7585

16   Isom LL (2002) Beta subunits: players in neuronal hyperexcitability? *Novartis Found Symp* 241: 124–138

17   Reenan RA, Hanrahan CJ, Barry G (2000) The mle(napts) RNA helicase mutation in drosophila results in a splicing catastrophe of the para Na⁺ channel transcript in a region of RNA editing. *Neuron* 1: 139–149

18   Hanrahan CJ, Palladino MJ, Ganetzky B, Reenan RA (2002) RNA editing of the Drosophila para Na(⁺) channel transcript. Evolutionary conservation and developmental regulation. *J Neurosci* 22: 5300–5309

19   Kawahara Y, Ito K, Sun H, Aizawa H, Kanazawa I, Kwak S (2004) Glutamate receptors: RNA editing and death of motor neurons. *Nature* 427: 801

20   Liu Z, Song W, Dong K (2004) Persistent tetrodotoxin-sensitive sodium current resulting from U-to-C RNA editing of an insect sodium channel. *Proc Natl Acad Sci USA* 32: 11862–11867

21   Song W, Liu Z, Tan J, Nomura Y, Dong K (2004) RNA editing generates tissue-specific sodium channels with distinct gating properties. *J Biol Chem* 279: 32554–32561

22   Tan J, Liu Z, Nomura Y, Goldin AL, Dong K (2002) Alternative splicing of an insect sodium channel gene generates pharmacologically distinct sodium channels. *J Neurosci* 22: 5300–5306

23   Plummer NW, McBurney MW, Meisler MH (1997) Alternative splicing of the sodium channel SCN8A predicts a truncated two-domain protein in fetal brain and non-neuronal cells. *J Biol Chem* 272: 24008–24015

24   Gustafson TA, Clevinger EC, O'Neill TJ, Yarowsky PJ, Krueger BK (1993) Mutually exclusive exon splicing of type III brain sodium channel alpha subunit RNA generates developmentally regulated isoforms in rat brain. *J Biol Chem* 268 18648–18653

25   Kerr NC, Holmes FE, Wynick D (2004) Novel isoforms of the sodium channels Na$_V$1.8 and Na$_V$1.5 are produced by a conserved mechanism in mouse and rat. *J Biol Chem* 279(23): 24826–24833

26   Akopian AN, Okuse K, Souslova V, England S, Ogata N, Wood JN (1999) Trans-splic-

ing of a voltage-gated sodium channel is regulated by nerve growth factor. *FEBS Lett* 445(1): 177–182

27 Zimmer T, Bollensdorff C, Haufe v, Birch-Hirschfeld E, Benndorf K (2002) Mouse heart Na+ channels: primary structure and function of two isoforms and alternatively spliced variants. *Am J Physiol Heart Circ Physiol* 282 H1007–1017

28 Makielski JC, Ye B, Valdivia CR, Pagel MD, Pu J, Tester DJ, Ackerman MJ (2003) A ubiquitous splice variant and a common polymorphism affect heterologous expression of recombinant human SCN5A heart sodium channels. *Circ Res* 31: 821–828

29 Donahue LM, Coates PW, Lee VH, Ippensen DC, Arze SE, Poduslo SE (2000) The cardiac sodium channel mRNA is expressed in the developing and adult rat and human brain. *Brain Res* 887: 335–343

30 Renganathan M, Dib-Hajj S, Waxman SG (2002) Na(v)1.5 underlies the "third TTX-R sodium current" in rat small DRG neurons. *Brain Res Mol Brain Res* 106: 70–82

31 Fauchier L, Babuty D, Cosnay P (2000) Epilepsy, Brugada syndrome and the risk of sudden unexpected death. *J Neurol* 247: 643–644

32 Buchner DA, Trudeau M, George AL Jr, Sprunger LK, Meisler MH (2003) High-resolution mapping of the sodium channel modifier Scnm1 on mouse chromosome 3 and identification of a 1.3-kb recombination hot spot. *Genomics* 4: 452–459

33 Blum R, Kafitz KW, Konnerth A (2002) Neurotrophin-evoked depolarization requires the sodium channel Na(V)1.9. *Nature* 419: 687–693

34 Kafitz KW, Rose CR, Thoenen H, Konnerth A (1999) Neurotrophin-evoked rapid excitation through TrkB receptors. *Nature* 401: 918–921

35 Craner MJ, Damarjian TG, Liu S, Hains BC, Lo AC, Black JA, Newcombe J, Cuzner ML, Waxman SG (2005) Sodium channels contribute to microglia/macrophage activation and function in EAE and MS. *Glia* 49: 220–229

36 Okuse K, Malik-Hall M, Baker MD, Poon WY, Kong H, Chao MV, Wood JN (2002) Annexin II light chain regulates sensory neuron-specific sodium channel expression. *Nature* 417: 653–656

37 Fitzgerald EM, Okuse K, Wood JN, Dolphin AC, Moss SJ (1999) cAMP- dependent phosphorylation of the tetrodotoxin-resistant voltage-dependent sodium channel SNS. *J Physiol* 516: 433–446

38 Vijayaragavan K, Boutjdir M, Chahine M (2004) Modulation of $Na_V1.7$ and $Na_V1.8$ peripheral nerve sodium channels by protein kinase A and protein kinase C. *J Neurophysiol* 91: 1556–1569

39 Herzog RI, Cummins TR, Ghassemi F, Dib-Hajj SD, Waxman SG (2003) Distinct repriming and closed-state inactivation kinetics of $Na_V1.6$ and $Na_V1.7$ sodium channels in mouse spinal sensory neurons. *J Physiol* 551: 741–750

40 Dina OA, McCarter GC, de Coupade C, Levine JD (2003) Role of the sensory neuron cytoskeleton in second messenger signaling for inflammatory pain. *Neuron* 39: 613–624

41 Safo P, Rosenbaum T, Shcherbatko A, Choi DY, Han E, Toledo-Aral JJ, Olivera BM, Brehm P, Mandel G (2000) Distinction among neuronal subtypes of voltage-activated sodium channels by mu-conotoxin PIIIA. *J Neurosci* 20: 76–80

42 West PJ, Bulaj G, Garrett JE, Olivera BM, Yoshikami D (2002) Mu-conotoxin SmIIIA, a potent inhibitor of tetrodotoxin-resistant sodium channels in amphibian sympathetic and sensory neurons. *Biochemistry* 41: 15388–15393

43 Keizer DW, West PJ, Lee EF, Yoshikami D, Olivera BM, Bulaj G, Norton RS (2003) Structural basis for tetrodotoxin-resistant sodium channel binding by mu-conotoxin SmIIIA. *J Biol Chem* 278: 46805–46813

44 Le Y, Sauer B (2001) Conditional gene knockout using Cre-recombinase. *Mol Biotechnol* 17: 269–275

45 Akopian AN, Souslova V, England S, Okuse K, Ogata N, Ure J, Smith A, Kerr BJ, McMahon SB, Boyce S et al (1999) The tetrodotoxin-resistant sodium channel SNS has a specialized function in pain pathways. *Nat Neurosci* 6: 541–548

46 Djouhri L, Fang X, Okuse K, Wood JN, Berry CM, Lawson SN (2003) The TTX-resistant sodium channel $Na_V1.8$ (SNS/PN3): expression and correlation with membrane properties in rat nociceptive primary afferent neurons. *J Physiol* 550: 739–752

47 Nassar MA, Stirling LC, Forlani G, Baker MD, Matthews EA, Dickenson AH, Wood JN (2004) Nociceptor-specific gene deletion reveals a major role for $Na_V1.7$ (PN1) in acute and inflammatory pain. *Proc Natl Acad Sci USA* 101: 12706–12711

48 Metzger D, Chambon P (2001) Site- and time-specific gene targeting in the mouse. *Methods* (1): 71–80

Index

The PIR-Series
Progress in Inflammation Research

Homepage: http://www.birkhauser.ch

Up-to-date information on the latest developments in the pathology, mechanisms and therapy of inflammatory disease are provided in this monograph series. Areas covered include vascular responses, skin inflammation, pain, neuroinflammation, arthritis cartilage and bone, airways inflammation and asthma, allergy, cytokines and inflammatory mediators, cell signalling, and recent advances in drug therapy. Each volume is edited by acknowledged experts providing succinct overviews on specific topics intended to inform and explain. The series is of interest to academic and industrial biomedical researchers, drug development personnel and rheumatologists, allergists, pathologists, dermatologists and other clinicians requiring regular scientific updates.

Available volumes:
T Cells in Arthritis, P. Miossec, W. van den Berg, G. Firestein (Editors), 1998
Chemokines and Skin, E. Kownatzki, J. Norgauer (Editors), 1998
Medicinal Fatty Acids, J. Kremer (Editor), 1998
Inducible Enzymes in the Inflammatory Response,
    D.A. Willoughby, A. Tomlinson (Editors), 1999
Cytokines in Severe Sepsis and Septic Shock, H. Redl, G. Schlag (Editors), 1999
Fatty Acids and Inflammatory Skin Diseases, J.-M. Schröder (Editor), 1999
Immunomodulatory Agents from Plants, H. Wagner (Editor), 1999
Cytokines and Pain, L. Watkins, S. Maier (Editors), 1999
In Vivo *Models of Inflammation*, D. Morgan, L. Marshall (Editors), 1999
Pain and Neurogenic Inflammation, S.D. Brain, P. Moore (Editors), 1999
Anti-Inflammatory Drugs in Asthma, A.P. Sampson, M.K. Church (Editors), 1999
Novel Inhibitors of Leukotrienes, G. Folco, B. Samuelsson, R.C. Murphy (Editors), 1999
Vascular Adhesion Molecules and Inflammation, J.D. Pearson (Editor), 1999
Metalloproteinases as Targets for Anti-Inflammatory Drugs,
    K.M.K. Bottomley, D. Bradshaw, J.S. Nixon (Editors), 1999
Free Radicals and Inflammation, P.G. Winyard, D.R. Blake, C.H. Evans (Editors), 1999
Gene Therapy in Inflammatory Diseases, C.H. Evans, P. Robbins (Editors), 2000
New Cytokines as Potential Drugs, S. K. Narula, R. Coffmann (Editors), 2000
High Throughput Screening for Novel Anti-inflammatories, M. Kahn (Editor), 2000
Immunology and Drug Therapy of Atopic Skin Diseases,
    C.A.F. Bruijnzeel-Komen, E.F. Knol (Editors), 2000
Novel Cytokine Inhibitors, G.A. Higgs, B. Henderson (Editors), 2000
Inflammatory Processes. Molecular Mechanisms and Therapeutic Opportunities,
    L.G. Letts, D.W. Morgan (Editors), 2000

Cellular Mechanisms in Airways Inflammation, C. Page, K. Banner, D. Spina (Editors), 2000
Inflammatory and Infectious Basis of Atherosclerosis, J.L. Mehta (Editor), 2001
Muscarinic Receptors in Airways Diseases, J. Zaagsma, H. Meurs, A.F. Roffel (Editors), 2001
TGF-β and Related Cytokines in Inflammation, S.N. Breit, S. Wahl (Editors), 2001
Nitric Oxide and Inflammation, D. Salvemini, T.R. Billiar, Y. Vodovotz (Editors), 2001
Neuroinflammatory Mechanisms in Alzheimer's Disease. Basic and Clinical Research,
    J. Rogers (Editor), 2001
Disease-modifying Therapy in Vasculitides,
    C.G.M. Kallenberg, J.W. Cohen Tervaert (Editors), 2001
Inflammation and Stroke, G.Z. Feuerstein (Editor), 2001
NMDA Antagonists as Potential Analgesic Drugs,
    D.J.S. Sirinathsinghji, R.G. Hill (Editors), 2002
Migraine: A Neuroinflammatory Disease? E.L.H. Spierings, M. Sanchez del Rio (Editors), 2002
Mechanisms and Mediators of Neuropathic pain, A.B. Malmberg, S.R. Chaplan (Editors),
    2002
Bone Morphogenetic Proteins. From Laboratory to Clinical Practice,
    S. Vukicevic, K.T. Sampath (Editors), 2002
The Hereditary Basis of Allergic Diseases, J. Holloway, S. Holgate (Editors), 2002
Inflammation and Cardiac Diseases, G.Z. Feuerstein, P. Libby, D.L. Mann (Editors), 2003
Mind over Matter – Regulation of Peripheral Inflammation by the CNS,
    M. Schäfer, C. Stein (Editors), 2003
Heat Shock Proteins and Inflammation, W. van Eden (Editor), 2003
Pharmacotherapy of Gastrointestinal Inflammation, A. Guglietta (Editor), 2004
Arachidonate Remodeling and Inflammation, A.N. Fonteh, R.L. Wykle (Editors), 2004
Recent Advances in Pathophysiology of COPD, P.J. Barnes, T.T. Hansel (Editors), 2004
Cytokines and Joint Injury, W.B. van den Berg, P. Miossec (Editors), 2004
Cancer and Inflammation, D.W. Morgan, U. Forssmann, M.T. Nakada (Editors), 2004
Bone Morphogenetic Proteins: Bone Regeneration and Beyond, S. Vukicevic, K.T. Sampath
    (Editors), 2004
Antibiotics as Anti-Inflammatory and Immunomodulatory Agents, B.K. Rubin, J. Tamaoki
    (Editors), 2005
Antirheumatic Therapy: Actions and Outcomes, R.O. Day, D.E. Furst, P.L.C.M. van Riel,
    B. Bresnihan (Editors), 2005
Regulatory T-Cells in Inflammation, L. Taams, A.N. Akbar, M.H.M Wauben (Editors), 2005